Image Processing and Pattern Recognition
Based on Parallel Shift Technology

Image Processing and Pattern Recognition
Based on Parallel Shift Technology

Stepan Bilan

State Economy and Technology University of Transport
Kiev, Ukraine

Sergey Yuzhakov

Gaisin United State Tax Inspection of the
Main Governance of SFS in Vinnytsa Region
Gaisin City, Ukraine

CRC Press
Taylor & Francis Group
Boca Raton   London   New York

CRC Press is an imprint of the
Taylor & Francis Group, **an informa** business

A SCIENCE PUBLISHERS BOOK

CRC Press
Taylor & Francis Group
6000 Broken Sound Parkway NW, Suite 300
Boca Raton, FL 33487-2742

Printed on acid-free paper
Version Date: 20171206

International Standard Book Number-13: 978-1-1387-1226-3 (Hardback)

Library of Congress Cataloging-in-Publication Data

Names: Bilan, Stepan, 1962- author. | Yuzhakov, Sergey, author.
Title: Image processing and pattern recognition based on parallel shift
    technology / Stepan Bilan, Ukraine, Kiev, State Economy and
    Technology University of Transport, Lesi Ukrainky str., 72, apt. 36, Kiev
    region, Vyshneve city, Ukraine, 08132, Sergey Yuzhakov, Ukraine, Gaisin
    united state tax inspection of the Main governance of SFS in Vinnytsa
    region, Heroes Krut str., 3, apt. 3, Vinnytsa region, Gaisin city,
    Ukraine, 23700.
Description: Boca Raton, FL : CRC Press, Taylor & Francis Group, [2018] | "A
    Science Publishers Book." | Includes bibliographical references and index.
Identifiers: LCCN 2017044421 | ISBN 9781138712263 (hardback : acid-free paper)
Subjects: LCSH: Computer vision. | Pattern recognition systems. | Parallel
    processing (Electronic computers)
Classification: LCC TA1634 .I495 2018 | DDC 006.6--dc23
LC record available at https://lccn.loc.gov/2017044421

Visit the Taylor & Francis Web site at
http://www.taylorandfrancis.com

and the CRC Press Web site at
http://www.crcpress.com

Preface

The development of computer image processing and recognition systems is one of the main aspects of artificial intelligence. Currently, there are no effective methods for video information manipulation other than digital image processing. A number of alternative methods and algorithms for processing images based on parallel-shear technology are proposed in this book. Using these along with digital image processing will enable the building of a hybrid image processing system. The methods and algorithms can also be used in analog computers.

The book describes images not in the form of an array of points, but as a set of functions. Methods and algorithms for image processing have been developed by analyzing these functions, using these it is possible to solve the main tasks of image processing. Description of images as a set of functions allows one to manipulate both raster and non-bitmap images. Only one quantitative characteristic is analyzed - the area of the image. The use of parallel shift technology also makes it possible to process images consisting of irregularly arranged elements. Scientific work is built in accordance with this classification.

The book is devoted to the development and research of methods of applying multiple transformations for computer processing and image recognition. Unlike the known methods, the proposed method uses a smaller number of the characteristic features to describe images and the number of computational operations in the system are reduced. Based on parallel shift technology, a mathematical model of the relationship between the form of the object and the functions of the intersection area, was developed. Methods of image recognition, determination of spatial orientation of objects, parameters of their movement and determination of the contour of a figure are described in the book. The study of the efficiency of pattern recognition with various parameters of the input image, are described.

The book is intended to help researchers in the development of artificial intelligence, robotics, software and hardware applications.

Contents

Preface   *v*
Introduction   *xi*

1. Feature Extraction Based on Parallel Image Shift Technologies   1

    1.1 Fundamentals of Understanding and of Constructing
        Images   1
    1.2 Image Description   2
    1.3 Characteristic Features of Images   5
    1.4 Theoretical Description of the Parallel Shift Technology   7

**2. Function of the Area of Intersection Formation Based
on Parallel Shift Technology   14**

    2.1 Mathematical Model of the Image Describing the
        Function of the Area of Intersection   14
    2.2 Analytical Determination of the Function of the Area of
        Intersection of Simple Shapes on the Basic Parameters   18
    2.3 Additional Parameters, which Accelerate Plane Image
        Recognition Process   25
    2.4 Algorithm of Recognition Images Based on PST   29
    2.5 Processing of 3D-Objects through PST   33

3. Methods of Pre-Processing Images Based on PST   35

    3.1 Edge Detection Method of Plane Figures   37
    3.2 Use of Parallel Shift Technology for Noise Removal
        in Images   40
    3.3 Technical Methods for FAI Preparation   47
    3.4 Model for Obtaining FAI of No Bitmap   49

4. Methods of Dynamic Images Processing Based on PST   50

    4.1 Determination of Spatial Orientation of Planes   52

4.2 Use of Different Types of Shifts of Image Copies in
    Video Processing Devices Based on PST   55

4.3 Methods for Determining Movement Parameters
    Using Cyclic Shifts   57

4.4 Methods for Determination of Motion Parameters
    Using Non-cyclic Shifts   59

4.5 Methods for Determining Distance of a Moving Object   61

4.6 Comparison of Methods for Determination of Motion
    Parameters by Using Different Types of Shifts of the
    Image Copy   63

5. Image Processing System Based on PST   **65**

5.1 System of Perception of Video Information   66

5.2 Saving Template Functions of the Area of Intersection   72

5.3 Application of Image-processing Methods Based on
    PST for Prosthetic Visual Organs   77

5.4 Perspective Directions of Image-processing Research
    Using PST   80

**6. Assessment of Productivity of the Image Processing and
Recognition Systems Based on FAI Analysis**   **84**

6.1 Evaluating the Efficiency of the Chosen Method of
    Removing Noise   85

6.2 Determination of the Permissible Error Range of Values
    Used to Limit the Maximum Displacement Values   90

6.3 Determination of Values Range of Acceptable Error
    Used in Comparison of Integral Coefficients   91

6.4 Determination of Range Values of Acceptable Error in
    Comparison of FAI on a Detailed Stage of the Analysis
    of Image Recognition   93

6.5 Evaluating the Effectiveness of Image Recognition
    Process in Various Parameters of Input Information
    Using Calculated Values of Permissible Errors   94

**7. Hardware Implementation of the Image Recognition
System Based on Parallel Shift Technology**   **101**

7.1 Generalised Model of Image Recognition System Based
    on Parallel Shift Technology   101

7.2 Selecting a Cover for Representing the Image in the
    Recognition System   102

7.3 Hardware Implementation of Parallel Shift and
    Computation of the Area of Intersection of Images   104

7.4 Modelling of the Main Hardware Modules for
    Determining FAI in Modern CAD Systems   113

7.5 Selection and Storage of Patterns in UMFP 118
7.6 Optical Image Recognition System Based on Parallel
  Shift Technology 120

**8. Methodology of Experimental Studies in Recognition Process
Based on Parallel Shift Technology 126**

8.1 Selection and Analysis of Characteristic Features of
  Images Based on Parallel Shift Technology 126
8.2 Analysis of the Accuracy of Image Recognition, Based
  on the Obtained FAI 131
8.3 Description of Image with the Help of Quantitative
  Characteristics of the Basic Properties of FAI 148
8.4 Experimental Analysis of FAI of Patterns Image 157
8.5 The Sequence of Steps for Using Parallel Shift
  Technology to Describe and Recognise Images 163
8.6 Advantages of Text Recognition Based on Parallel
  Shift Technology 166

References 187
Index 192

Introduction

Today, we make extensive use of advanced information technology. One of the main problems faced by the scientific community and industry is to create new and more sophisticated tools to help in investigation. The development of computer technology made it possible to carry out a major step forward in the field of information processing. The next step of progress in this direction was the development of robotics. The need to process large volumes of information is encouraged by the creation of powerful intelligent systems that are based on modern achievements of computer technology. Intelligent Systems (IS) are necessary to replace the person in the processes that entail significant physical, mental and emotional strains. One of the criteria to improve performance of IS is the availability of effective receptors (Tou and Gonzalez, 1977).

Intelligent systems are built according to the scheme: receptor—analysis unit—effector. The most developed intellect is present in biological creatures, which have the best parameters of these components.

Today we know that a lot of information about the world can be obtained by analysing the video data. For example, a person perceives 90 per cent of information by using the photoreceptors. Using these facts, we can conclude that in the construction of intelligent systems, the need is for creation of the means of information perception.

To perform the tasks of processing video the modern man has effective receptors—the eyes. In most living beings, the eyes have a similar structure, and accordingly, video data processing passes through the same principles. The developers of machine vision systems are trying to achieve, and eventually surpass, the level of quality and efficiency of image processing.

Opportunities of clustering and further classification of objects directly depend on the number of characteristic features (CF) (Shalkoff, 1989) and the appropriateness of their use in solving one or other image processing tasks. The more the characteristic features are selected from images, then the total number can be divided into a greater number of classes. At this stage, the development of intelligent systems is restricted by the absence of unambiguous methods for the selection of the characteristic signs (Chen et al., 1995). All CF are defined by the system designer based on experience in this work.

To visualise images, the characteristic features are obtained by analysis of their geometrical properties (edge detection, search areas of connectivity, calculation of the number of angles and others.).For hardware implementation, these processes require the execution of various algorithms based on a large number of mathematical operations that can be quite complex. This increases the time spent on data processing.

One possibility to improve the performance of image preprocessing is to use a little amount of homogeneous characteristic features of object that would be formed by using a simple algorithm without loss in quality of information processing. We propose to use the simple physical quantity—the area of the image. This value in various multiple values of quantities can represent and describe various images that have different shapes and belong to different classes. For transformation of the area into a set of characteristic features it is necessary to perform dynamic changes using the parallel shift of images.

The reason for this proposal is the fact that the human eye is composed of 120-150 million photoreceptors, and the thickness of the optic canal is estimated at about 1 million axons. In the processing of the video signal is used a specific grouping of information. In this book the image processing methods are shown that use parallel shift technology. Use of this technology will allow replacement of the set of a large number of characteristic features of the image, which are formed by different algorithms, sometimes with quite complex mathematical operations and homogeneous characteristic feature (area) (Winston, 1992). This feature is measured and is calculated by using simple circuit solutions.

At this stage of development of machine vision systems a parallel shift process is considered by researchers solely for the purpose of normalising the image for further recognition. That is, they are used in their algorithms and affine transformation only for replacement of the input image to another, more suitable for further processing.

The main hypothesis of this work is the use of parallel shift to obtain a set of image properties, which allows its further processing and analysis of those or other recognition methods. Adding the dynamics of change, the static parameter of the image as the area will allow it to be used for classification and definition of the image parameters. The image is transformed into a set of functions that allows processing of any means of functional analysis. The components of these functions will form a set of CF images for further analysis.

Now there are no methods that serve as alternatives to the digital image processing. Many methods and means of digital image processing have been developed. The quality of their performance increases with increase of power of a computing means. However, this quality is based on extensive increase of the resolving capacity of devices. A qualitative change is necessary for further progress.

As shown below, the use of parallel shift technology (PST) allows the performance of all the basic processing tasks. Some of these tasks with the help of PST are easier to implement than using digital image processing. To improve the performance of the machine vision, it is advisable to build hybrid systems that will combine elements of digital processing and parallel shift technology. Such machine vision systems are especially necessary for construction of autonomous robots. The vision systems based on PST can be used for rapid decision-making. Digital image processing is necessary for the detailed analysis of images.

The image processing system based on parallel shift technology allows the processing of images. To enable such processing, it is necessary to create a means for determining the area of intersection of the figure and its copy that moves in parallel. The image processing system based on PST can be used in analog computers.

The book presents the material in eight chapters. The problems of image processing and pattern recognition theory based on parallel shift technology are considered. Methods are described for image pre-processing, such as edge detection, noise removal, dynamic images processing, etc. The book also displays the results of experiments based on the developed software models that confirm the effectiveness of the developed models and methods.

Feature Extraction Based on Parallel Image Shift Technologies

1.1 Fundamentals of Understanding and of Constructing Images

We perceive the information through our senses. One of the main sources of information for our brain is the human visual channel. All of us have the visual channel with an identical structure. There are differences in the characteristics of the channels determined by age, or a variety of pathological changes caused by diseases, or external influences, etc. However, the purpose of human visual channel is to transform the optical signals of the visible spectrum into a set of electrical or chemical signals that are received by the optic nerve in the region of the human brain, which is responsible for processing of visual images.

The visual images in the surrounding world are constantly changing our photoreceptors (the bulbs, the sticks). The visual image is a set of optical signals that fall on the field of our primary receptors. This set of optical signals present the visual picture.

When the optical picture falls on a medium that can record it, we use the term image. The physical medium, which displays the recorded optical image is called the media of the image (paper, storage media, optical/electronic media, etc.).

The image may be superimposed on the carrier in two ways:

- Fixation of the optical image by designing and memorising by using the physical storage materials or via opto-electronic storage elements.
- Construction and reconstruction of images by using special graphic means.

In the first method, complete copying is carried out of the optical pattern on the carrier. The accuracy of the copying is determined by the structure of the medium on which the image is projected. The second method uses methods which are based on full knowledge of image information and reconstruction of it on a carrier. Both the methods require the construction and use of special hardware. The first method does not require any specific knowledge about image structure. The optical pattern can be reproduced on the medium in parallel (picture, optoelectronic matrix and the like) or successively with different scanning means. The image is formed by an exact repetition of colour and brightness elements of the image. Such elements can be points or a specific areas of the specified shape image and size. The points have a minimal area and the form elements can have a large area. Such elements can be triangles, rectangles and various predefined geometric components of complex images. The second method always takes into account the adopted structure of the image and is characterised by a set of methods that are implemented on the developed hardware/software tools.

This model of the image is created for the hardware that can implement it and reproduce the image on the selected medium. The model describes the image. The complexity of the hardware and the accuracy of the reconstruction depend on the chosen model. Choosing the model is one of the main tasks when constructing an image for modern processing and image recognition.

In modern data processing systems, there are several basic tasks that can be divided into tasks of reproduction images and image understanding. Image-understanding tasks consist of a preliminary description of the image. Description of images is based on the image representation of functions and by a set of elementary (discrete) components (e.g., pixels, cells, etc.) that are most effective in further processing.

The image understanding task is to analysis of a geometric form of the image and the successful allocation of the necessary elementary components of the image and the links between them or the functions with which the signification of image definition can be obtained.

These two tasks are opposites and occupy a major place in the construction of intelligent systems.

1.2 Image Description

Images can be divided into bitmap and non-bitmap images (Nixon and Aguardo, 2002; Solomon and Breckon, 2011).

A bitmap is an image that includes a set of pixels (usually rectangular) on the opto-electronic display device or on paper and other means. Each pixel can display a different colour. Bitmap images are now most commonly used in most input-output devices. Non-bitmap image is an image that is not described in pixels but is seen in elementary geometric objects, called primitives. In most cases, the primitive are mathematical models.

Most often non-bitmap images are ascribed as vector images. Vector images are processed by using the same computational tools as that for bitmaps. They can be reproduced on both raster and vector on the image display device. However, raster images can only be displayed on raster reproduction devices.

The vector-constructing image method faces problems of building complex images, since it requires storage of a large amount of primitives and their mathematical description. Vector images scale well without loss of information while bitmap images are scaled poorly. In the bitmap image after scaling, a large number of pixels is lost (for size reduction) and appears in increased size.

In the optical system, an image is used as a unified optical object. The images in such systems are transformed by optical-mechanical means and can be divided into individual components. By using special optical-mechanical means, the whole image can be displayed on the selected medium. However, in such systems there is the problem of storing images and of efficient software manipulation. At the same time, optical systems have a high speed and accuracy of images processing. Currently, optical image processing and recognition systems are increasingly used in various fields of human activities. In addition, they do not use the methods of computer processing and recognition. Such systems work with analog images and use specialised imaging methods (Pratt, 2016).

In the modern computer systems of image displaying and image recognition, all the methods are implemented on the basis of the hardware and software. The most effective description images method is to use the structural describe (Solomon and Breckon, 2011; Pratt 2016). In some cases, the image is convenient to represent by topological description (Solomon and Breckon, 2011; Pratt 2016).

The simplest way to describe the images, which are fed by a certain electron carrier, is to describe them by a set of discrete elements and function of the brightness in three-dimensional or two-dimensional space (Nixon and Aguardo, 2002). This function describes the value of the average brightness of the image in the neighbourhood of each discrete element (cell or pixel). The brightness function is discretised and determines the gradation number, on which is divided its range of values. For binary and grey-scale image, the gradation number is equal to or more than two. Based on this method are implemented the cost-effective methods of describing, which constitute the basis for other methods. An important requirement and limitation are the size and structure of the set of describing elements.

In many problems it is enough to use a simplified description, which consist of selection and analysis of the image contour (Nixon and

Aguardo, 2002; Belan, 2011; Kozhemyako et al., 2001; Bilan, 2014). To date, a range of edge detection methods were offered and morphological processing methods (Solomon and Breckon, 2011; Parker 2010). The main disadvantage of this method is that any calculation error depends on the parameters situation, which affects the accuracy of the characteristic elements.

Simple images are more cost-effective in describing and identifying objects, as well as in dividing into elementary shapes that are determined by approximation of the original image (Solomon and Breckon, 2011; Pratt, 2016; Gonzalez and Woods, 2008).

One of the methods of describing complex images is a structural description. The method is based on the representation of images as a structure consisting of elementary components (terminals) (Narasimhan, 1966; Stallings, 1972; Shaw, 1969; Hancock et al., 2010; Gimel'farb et al., 2012). These include the syntactic description method that consists of selection of a set of terminals that constitute nominal features. A set of non-terminal symbols is being introduced, among which the first is determined. The rules for passing character chains with terminal and non-terminal symbols are defined by a set of rules of substitutions. The terminal and non-terminal symbols and substitution rules create a formal grammar. Options of syntactic approach are defined as a specific type of terminal elements and by complexity of the rules of substitutions. The usual syntactic approach to the description is same, but the loosely oriented images do not give the same grammar, that is, they make changes to their description.

The imaging algorithms are based on the method of potential functions (Aizerman, 1970), where the "elementary function of closeness" between two vectors is introduced and this is called the potential function. The proximity of the vectors is determined and its maximisation is used, which is a serial exhaustive search for all the vectors of the set. Also attempts are made to use other coordinate systems that describe the image elements.

Often images in the topology are described that characterize the general properties of the image (the presence of holes or areas). These properties remain for various common distortions. The most common topological feature is Euler's number G (Sossa-Azuela, 2013; Acharya, 2006; Bishnu, 2005), which is defined by the formula

$$G = C - H,$$

where C – the number of associated image components

   H – the number of holes

In the tasks of reproduction and image recognition the most common method to describe images is use of a technique that is based on the formation of vector of a geometry, brightness and colour features. These characteristics include the area, the perimeter elements, and the moment functionals from the brightness function, for example, the centre of gravity of the image (Maier, 2012; van Assen, 2002).

The information representation of the image is its spectral representation. The definitions of contour characteristics (the function of the polar φ angle, contour length, the distance between the centre of gravity and the contour points) are used. The curvature of the contour is expanded in a Fourier series and the factors are defined to make it possible to describe the image (Nixon and Aguardo, 2002; Tremeau and Borel, 1997).

A successful description and presentation of images greatly affects the efficiency of their recognition as well as display.

1.3 Characteristic Features of Images

The characteristic features describe the image by using the minimum set of elements with minimal loss of image restoration. The characteristic features may be set in advance, or extracted from the image for its recognition.

The recognition task is used when the input image is unknown. In this case, the characteristic features are extracted by an image recognition system itself. Recognition is characterised by a process that assigns an object to one of the previously specified fixed list (class) according to a specific set of rules in accordance with the intended purpose. Recognition can be carried out by any recognition system that generates a set of features, according to which the system determines the class to which the image belongs.

Extracted characteristic features of the image at the input by a system of recognition provide a meaningful description of the image.

However, at the moment there is no universal method for selection of a full set of characteristic features for all the images. For each class of images, a set of features is selected by developers of the recognition system. These characteristic features do not always give 100 per cent results and cannot be applicable to other classes of images. For example, the characteristic features of the face of a person (the distance between the eyes, nose length, the shape of the lips and so on) cannot be applicable for human fingerprints (control points, lines form and so on) and conversely. Moreover, all these features are set in advance by the developer based on his intuitive experience and a thorough study of the selected class of images.

Today we can confidently assert that there are no methods and hardware that allow extraction of most characteristic features of the image by the recognition system itself. For example, the modern biometric identification system accurately enough recognises a face but caricatures of the face cannot be recognised. At the same time, the human visual channel recognises faces with distortions as well as its caricatures.

Developers typically solve special problems for image recognition with specific forms and brightness characteristics. Studies show that the greater the characteristic features extracted from the image the more accurate is the recognition result.

Today, a number of methods have been developed that are used to extract the characteristic features images (Nixon and Aguardo, 2002; Pratt, 2016). All these methods are based on preliminary image transformation. Among such transformations are Fourier transformation (Nixon and Aguardo, 2002; Bracewell, 1986; Solomon and Breckon, 2011), wavelet transformation (Nixon and Aguardo, 2002; Daubechies, 1990), Hough transformation (Gonzalez and Woods, 2008; Ballard, 1981), affine transformation (Belan, 2011; Bilan, 2014), discrete cosine transformation (Ahmed et al., 1974), Hartley transform (Bracewell, 1984a, Bracewell, 1984b) and other methods.

Different types of filtration are used for extraction of the characteristic features, which are fully described in various literatures (Nixon and Aguardo, 2002; Gonzalez and Woods, 2008). All of these approaches focus on extraction of the previously defined characteristic features or identifies those features that can extract by using these methods. Thanks to these methods the number of characteristic features may prove excessive and may not provide a complete description of the image.

Widespread image recognition methods do not extract the specific characteristic features. These include methods based on the use of artificial neural networks (ANN) (Xingui andShaohua, 2010; ICANN, 2014), cluster analysis (Anil K. and Jain, 2010; Chang et al., 2012; Berklim, 2002; Islam and Ahmed, 2013) and etc. require preliminary preparation of images for their recognition.

ANN uses a large database of patterns for recognition and require larger neural network learning costs. The most universal ANN include cognitron and neocognitron (Fukushima, 1983).

Cluster analysis is to find and combine image elements through specified properties. These properties can be shapes, colours, etc. Cluster analysis is very effective in image segmentation; genetic algorithms and fuzzy sets (Tahmasebi and Harakhani, 2012), and fractal theory (Barnsley, 1988, Saupe, 1996) are also used.

All these methods have substantial differences, but certain advantages and disadvantages. The most common approach is to extract the characteristic features of an image by using known or developed methods and the formation of the vector of the characteristic features. The extracted characteristic features may be rigidly located in the vector or may be independent of the location. Created vector in the memory of patterns is recorded and the identifier belonging to the class is assigned.

However, a created vector of the characteristic features does not always give the desired result. Also, the more the characteristic features are the more the computational operations and transformations the system must perform. This reduces the accuracy and speed of the recognition system.

In modern literature, almost no attention is paid to reducing the number of characteristic features. Authors did not know image processing and pattern recognition systems, which, on the basis of one or more characteristic features, would render it fully. These features need not necessarily be part of the image but may be the result of a quantitative indicator of the input image transformations.

According to the authors, the most simple image transformation is its shift and a quick and easy computable setting of the image is an area. The combination of the operation of parallel shift and calculation of the quantitative value of the image area gives good results in image recognition systems. In addition, the shift of images and the definition of its area are easily implemented in optical systems that work with analog images but the result may not be in digital form.

1.4 Theoretical Description of the Parallel Shift Technology

Parallel shift technology (PST) - the process organisation of interaction of the object, the elements of which are described by n-dimension vectors of the characteristic features, with its own copy, in elements which are changing only one or more CF than in the original (Belan and Yuzhakov, 2013). This process is similar to parallel shift in the space of variable characteristic features. To provide a process to the parallel shift, it is necessary to choose the characteristic features which have numerical values, for example, coordinates, time or phase.

Each element of the initial set is obtained by superposition pair of vectors with characteristic features, whereby the second element will be the element of parallel shifted set. The result of the interaction of vectors differs from initial vector characteristic features of the object and can be used for further analysis. The options available for analysis of the interaction of the original object and its copies can be any mathematical operations and the tools of analysing the sets. If the result obtained by parallel shift of a particular set of quantitative values for the different shift values, a set can be considered as a function of displacement. For example, there is an input object representing a sinusoidal signal. When shifting the sinusoidal signal-copy in phase space, its sum with an initial signal will be different from the sine wave and when the phase shift is by π it will be zero.

The use of parallel shift technology is advantageous when solving image recognition tasks and processing of dynamic objects.

In order to describe the technology of parallel shift in terms of set theory, we should consider the processes of clustering and classification (Duran and Odell, 1974, Islam and Ahmed, 2013). These processes are necessary for combining the elements of a set for selection of its subsets, which are similar by their certain characteristic features. The ability to distinguish one object from the other gives you the opportunity for further analysis.

Clustering is characterised by several factors, which include the number of clusters that compose the original set. At a constant amount of clusters, this process is called 'clear clustering' and in an uncertain amount of clusters, it is called 'fuzzy clustering'. In the case of clear clustering, each element of the original set belongs to only one subclass of elements. Subclasses of elements do not intersect. In fuzzy clustering, the subclasses of elements of the original set can intersect. Belonging of elements to a certain subclass is determined by some factor. This ratio is called degree of membership of object to the cluster. For elements with clear clustering, it takes the values 0 or 1, and for fuzzy clustering, it is in the range of 0 to 1.

Also the parameter of clustering is a measure of the distance. This parameter is selected to perform certain tasks randomly of some number-designed measures; for example, the most widespread methods for determining measures of distances are to use the Euclidean distance or Euclidean square distance. Using other measures of distances is determined by the need to allocate some feature of the set elements.

The third parameter is the clustering criterion. The choice of these criteria is based on the needs of the specific problem to be solved and on the developer's experience. The most common ones are determination of the maximum and minimum values specified in the characteristic features of the set element.

The basis for implementation of the clustering process is formation of the vector of characteristic features. This vector is generated to describe the object. The characteristic features can be both qualitative and quantitative. The composition of the set depends on the specific problem to be solved. The description of the variety of elements of vectors leads to complication of clustering algorithms. The use of parallel shift technology allows formation of a homogeneous characteristic features to describe the object.

Assume there is a set of objects x, which are located in the examined area of features. Each of them has some predetermined amount of N of characteristics.

$$\overline{x} = (CF_1,.., CF_N) \tag{1.1}$$

Parallel shift technology involves the selection of a total of one or more of the same type of quantitative features for each object x to form a transformed object xt.

$$(CF_i,.., CF_j) \in (CF_1,.., CF_N) \tag{1.2}$$

The list of CF is determined by the ability to describe the original object for solving a specific task; other features are discarded as not being essential in its decision.

$$\overline{xt} = (CF_1,.., CF_N) \tag{1.3}$$

The scheme for the process of conversion of x set by reducing the number of characteristic features in the vectors that describe its elements is shown on Fig. 1.1.

Thus, each of the original objects x is converted into some transformed object xt. The x and xt sets are bijective. Further manipulation is necessary to carry out the objects of the xt set.

We describe the process of using parallel shift technology in the case of discretized initial set. For example, for 2D-objects such a set can be considered as the digitized image.

In this case xt object is a set of m elements xt_i, each of which is described by a vector of the n characteristic features.

$$\overline{xt_i} = (cf_1,.., cf_N) \tag{1.4}$$

These features form the CFs, which are chosen to describe the transformed xt object and their number is same.

$$CF_j = \frac{\sum\limits_{i=1}^{m} cf_{i,j}}{m} \tag{1.5}$$

As a result of the interaction of the xt object and of xt' object, in which the value of all the characteristic features $(CF_i,.., CF_j)$ are changed in a certain shift direction in the space of selected characteristic features at a value SHIFT, a x_{pst} set. xt' object is obtained and it is a moved copy of the xt object.

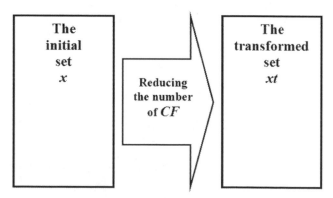

Fig. 1.1: Scheme of the set transformation process by reducing the number of characteristic features in the vectors, which describe the elements

$$CF'_j = CE_j + \text{SHIFT}_j = \frac{\sum\limits_{i=1}^{m}(cf_{i,j} + shift_j)}{m}. \tag{1.6}$$

The value SHIFT displacement will depend on the *shift_j* and displacement of each characteristic feature. If the shift is parallel, then $SHIFT_j = shift_j$; else, it is object deformation.

$$\text{SHIFT} = \sqrt{\sum\limits_{j=1}^{n}(SHIFT_j)^2} = \sqrt{\sum\limits_{j=1}^{n}\left(\frac{\sum\limits_{i=1}^{m}shift_{i,j}}{m}\right)^2}. \tag{1.7}$$

Variants of the organisation of interaction of the sets can be varied. In this book it is considered the simplest option-determination of intersection of two sets.

$$xt \cap xt' = x_{pst} \tag{1.8}$$

The scheme of the x_{pst} sets formation for some shifting distance SHIFT is shown in Fig. 1.2.

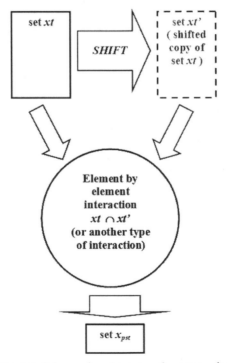

Fig. 1.2: Scheme for obtaining the set x_{pst} for some distance of displacement

The x_{pst} set is only a part of the xt set, so information about its elements is not enough to assign the initial x element to a certain cluster. A x_{pst} set can determine its cardinality (crd). For 2D-objects, this is the area of intersection of images; for 3D-objects, this is the volume of intersection, etc. These parameters are numerical values, so it is not enough to describe the x elements. It is necessary to make the dynamics of change of these static parameters. For each possible direction of the shift in space of the selected characteristic features ($CF_i,.., CF_j$), each possible SHIFT shift distance can be determined by the crd parameter. Using a set of all possible crd parameters, each x element can be described. For each direction of parallel shift, a sequence of crd parameters will be a function of the x_{pst} set cardinality that depends on the shift. The set of all such functions for each element x is used for further processing and analysis by a variety of methods based on parallel shift technology.

Assume the average value of cardinality of the elements of x_{pst} set, which belong to one cluster, for each shifting distance $SHIFT=j$ will be equal to CRD_{SHIFTj}. The number of elements in the specified i-cluster will be cnt_clast_i. They are related by the following relation:

$$CRD_{SHIFT_j} = \frac{\sum_{i=0}^{cnt\_clast_i} crd_{SHIFT_{i,j}}}{cnt\_clast_i} . \tag{1.9}$$

For comparison functions in the process of clustering, we introduce a threshold value trh. The criterion for clustering of objects will be assumed and the deviation values of the function is $crd_{SHIFT_{i,j}}$ and depend on the CRD_{SHIFTj} for the i-th cluster as each shift distance $SHIFT = j$ will not more than the threshold value.

$$\left| CRD_{SHIFTj} - crd_{SHIFT_{i,j}} \right| \le trh . \tag{1.10}$$

Schematic representation of the CRD function of the mean value cardinality of the x_{pst} set and the range of possible crd value features that belong to one cluster is selected in grey in Fig. 1.3.

In Fig. 1.3, the CRD graph is shown for one direction of displacement of the xt' set in the space of selected CF. For a description of the xt object it is essential to use the set of functions for all shift directions. It should be noted that this example gives fixed value of trh threshold for all possible displacement range. The deviation value of threshold can be changed, depending on the shift distance. Due to possible changes in trh parameter, the belonging of object to one or another cluster will also change. This fact confirms the principles of 'fuzzy clustering' of the set.

To describe objects of studies by using a set of functions, any methods of functional analysis can be used. Here dualism is manifested similar

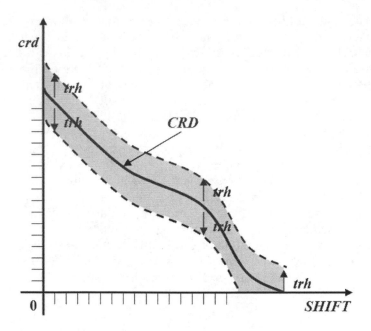

Fig. 1.3: Schematic representation of the CRD function of the mean value cardinality of the x_{pst} set and the range of possible *crd* value features that belong to one cluster is selected in grey

to the wave-particle theory. Sets in the functions are transformed. The function parameters can be combined *cfc* factors which themselves can be used to describe the initial set.

$$\overline{x} = (cfc_1, ...cfc_{angle}) \tag{1.11}$$

where *angle* is the number of possible shift directions.

For objects that are presented by irregularly positioned elements or do not have a clear structure, it is not possible to express the initial set through a set of elements. In this case, it is necessary to have the available means to fix the result of interaction between the original object and it is copy at the place of intersection. In other respects, the principles of further analysis of the obtained functions are the same. This is one of the main advantages of using the parallel shift technology.

This scientific work is devoted to the use of parallel shift technology for image processing. Further, all the above will be described for plane images processing that are transformed to a set of functions.

To obtain functions of the area of intersection of images the change of the object's coordinates is used. The superposition result of the original image and its parallel shift copies is the area of intersection field of this objects. The set of values of area intersection for each moment of change of coordinate constitutes the function of the area of intersection.

The principles of transformation (Ulam, 1952) of objects and information processing outlined in this book can be applied in other industries that differ from image processing.

Function of the Area of Intersection Formation Based on Parallel Shift Technology

2.1 Mathematical Model of the Image Describing the Function of the Area of Intersection

One of the applications of parallel shift technology is in the imaging processes. Currently, there aren't any effective methods for handling video data than digital image processing. The quality of digital images increases with the increase of resolution capability of the devices. At the same time, the data size of the array that describes the image increases too. Also, the required number of operations to process the array goes up.

It is necessary to reduce the number of elements, which describe the image. Among the attributes of the image, the one that has a quantitative expression is the area. By itself, the area contains information only about the size of the image. The use of parallel shift technology for image processing allow to form a set of functions for its description. A parallel shift in this case is the change of the spatial coordinates. Analysis of the obtained set of functions can be the base for performing basic image-processing tasks.

The function of the area of intersection (FAI) depends on the area of intersection of the original image and its copy, which parallely shifts in the selected direction, as well as on a distance of shifting (Belan and Yuzhakov, 2013).

FAIφ is defined as the intersection area of figures A and its copy B, which is shifted in the direction parallel to φ (Fig. 2.1a). The intersection area at each time point is denoted as C:

$$A \cap B = C$$

The shift is made from the position of full match figures A and B copy till the moment of zero intersection. The total distance over which a copy of the image moves till the moment of zero intersection is called the maximum shift (X_{max}). In this process, the static characteristic of the image (area) turns into a function, which can be analysed to identify the figure and its parameters (Fig. 2.1b).

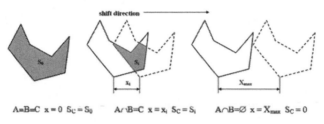

$$A{=}B{=}C \quad x = 0 \quad S_C = S_0 \qquad A{\cap}B{=}C \quad x = x_i \quad S_C = S_i \qquad A{\cap}B{=}\varnothing \quad x = X_{max} \quad S_C = 0$$

a) stages intersection of image and its copy

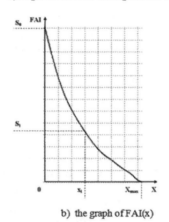

b) the graph of FAI(x)

Fig. 2.1: The model obtaining FAI(x)

The function of the area of intersection that is obtained by this model will be the same for the class of figures (X), which differ from each other only by mutual position of the components arranged in the displacement direction (Fig. 2.2).

In order to distinguish between images that belong to the class with the same FAI, we must carry out the definition of at least another one FAI for the direction $\varphi_1 \neq \varphi$. Thus Y class figures are obtained which have the same FAI. The intersection of classes of Figures X and Y corresponds to the class F1 figure, which includes the recognisable image.

$$X \cap Y = F_1.$$

The function of the area of intersection may be determined for more directions. Increasing the number of FAI for describing the image reduces

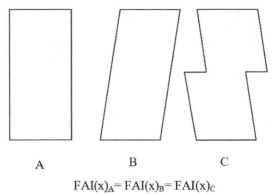

A   B   C

$$\text{FAI(x)}_A = \text{FAI(x)}_B = \text{FAI(x)}_C$$

Fig. 2.2: Figures with equal FAI(x)

the number of elements in class F2 figures, which includes the original image.

$$X_1 \cap X_2 \cap ... \cap X_i \cap ... \cap X_n = F_2$$

where X_i – figures class with the same FAI towards i.

$$F_1 \geq F_2 \geq 1$$

Use of the above properties helps to develop a mathematical model of the image representation, which help the function of the area of intersection. To identify the image more than one FAI is used and that's why the shift will be in two orthogonal directions. Orthogonal directions are chosen in order to simplify the software implementation.

Input Fig. 2.3a is bound above by a function $f_2(x)$ and below by a function $f_1(x)$. The image is aligned with the lower edge of Fig. 2.3a. It is bound below by an axis 0X and on top by a function $f_3(x)$, where

$$f_3(x) = f_2(x) - f_1(x) \tag{2.1}$$

Since image data differ only in the mutual position of components, they are arranged in the shift direction (in this case, vertical); that's why $\text{FAI(}y\text{)}_a = \text{FAI(}y\text{)}_b$. The possibility of describing an object with parameters is reflected in the formula (2.1). These parameters are derived from the parallel shift technology. Obviously, the function $f_3(x)$ that is obtained as a superposition of functions $f_2(x)$ and $f_1(x)$, consists of contour of the object. This property allows its use for study of the form shapes.

$$FAI(y) = \int_{x_1}^{x_2} f_3(x)\,dx - y \cdot (x_2 - x_1). \tag{2.2}$$

The values of x_1 and x_2 are determined from the system (2.3).

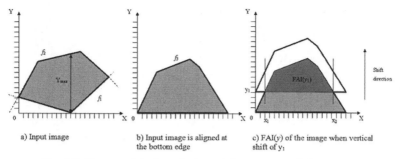

a) Input image

b) Input image is aligned at the bottom edge

c) FAI(y) of the image when vertical shift of y_1

Fig. 2.3: The graphic description of the process of obtaining a mathematical model of construction of the FAI(y)

$$\begin{cases} y = f_3(x_1) \\ y = f_3(x_1)' \end{cases}$$ (2.3)

where y is offset value at this stage.

The analogous method (Fig. 2.4) is used to determine the FAI(x), but with the difference that the displacement is carried out in the other direction

$$f_6(x) = f_5(x) - f_4(x)$$ (2.4)

$$FAI(x) = \int_{y_1}^{y_2} f_6(y)\,dy - x \cdot (y_2 - y_1)$$ (2.5)

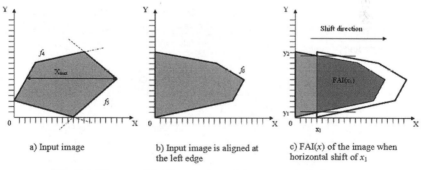

a) Input image

b) Input image is aligned at the left edge

c) FAI(x) of the image when horizontal shift of x_1

Fig. 2.4: The graphic description of the process of obtaining a mathematical model of the construction FAI (x)

The values y_1 and y_2 are determined from the system (2.6).

$$\begin{cases} x = f_6(y_1) \\ x = f_6(y_2) \end{cases}$$ (2.6)

where x is offset value at this stage.

As can be seen from Figs 2.3 and 2.4 the parts of the functions $f_2(x)$ and $f_1(x)$, on which is based the definition of FAI, contains the same image points as functions $f_5(y)$ and $f_4(y)$. This property provides redundancy that can be used for recognition.

If the function of the area of intersection is defined for the figures, which are not convex, or have any foreign objects within their images, it is necessary to take account of such changes inform. As can be seen from above, the FAI is built on functions that reflect the shape of the figure. They themselves become more complex with increasing complexity of image shapes, so their elements can be used as characteristic features in the process of image description.

The basic parameters of the function of the area of intersection for describing images by two FAI is the initial image area (S_0), the maximum shift in the direction of φ (X_{max}) and the maximum shift in the direction of φ_1 (Y_{max}). When images are described by the larger number of FAI, the maximum shift for each direction φ_i (I_{max}) is determined.

The maximum shift is the minimum distance of shift in the image copy in the chosen direction, where there is no intersection with the original image. Only on the basis of these parameters, further calculations are performed. Determination of these parameters is easy to implement in hardware. Especially effective is the use of cellular automata (CA) (Belan and Belan, 2012; Belan and Belan, 2013; Bilan, 2014) to build a matrix of homogeneous environments, which implement parallel shift of data sets and determination of the area.

2.2 Analytical Determination of the Function of the Area of Intersection of Simple Shapes on the Basic Parameters

The figures, which have the function of the area of intersection that can be determined analytically, the simple figures (SF) can be called (Yuen et al., 1989). A triangle, square, rectangle and circle belong to the group of simple shapes. Besides, we can analytically determine FAI for any regular $2n$-setsquare. However, if n increases, then the values of these functions will rapidly approach the FAI of the circle. Furthermore, the function of the area of intersection requires processing by dividing into the displacement ranges depending on n.

Examples of hexagon image conversion process at FAI formation for different angles of inclination to the horizontal are shown in Fig. 2.5.

For different areas of the shift different copies are used for the functions of the area of intersection. Two sections defining a function of the area of intersection are used for the hexagon. In Fig. 2.5, they correspond to distance shifts of image copies which equal l_1 and $l_2 - l_1$. In general, the

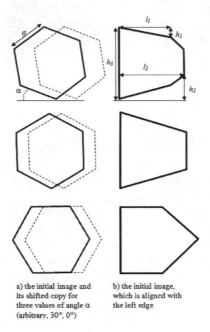

a) the initial image and
its shifted copy for
three values of angle α
(arbitrary, 30°, 0°)

b) the initial image,
which is aligned with
the left edge

Fig. 2.5: Sample images of the hexagon in the FAI formation process

parameters of the alignment on the left-edge hexagon will be as following:

$$h_1 = a \sin \alpha \tag{2.7}$$

$$h_2 = \frac{a}{2}\left(\sqrt{3}\cos\alpha - \sin\alpha\right) \tag{2.8}$$

$$h_3 = 2a \sin\left(\frac{\pi}{3} + \alpha\right) = a\left(\sqrt{3}\cos\alpha + \sin\alpha\right) \tag{2.9}$$

$$l_1 = \frac{\sqrt{3}a}{\sqrt{3}\cos\alpha - \sin\alpha} \tag{2.10}$$

$$l_2 = \frac{2\sqrt{3}a}{\sqrt{3}\sin\alpha + \cos\alpha} = X_{max} \tag{2.11}$$

When identifying the input image as a hexagon, we can determine the length of its sides using the basic parameters of the function of the area of intersection.

$$\begin{cases} a = \sqrt{\dfrac{2S_0}{3\sqrt{3}}} \\[2mm] a = \dfrac{X_{max}}{\sqrt{3}}, \text{for } \alpha = \dfrac{\pi}{6} \\[2mm] a = \dfrac{X_{max}}{2}, \text{for } \alpha = 0 \end{cases} \qquad (2.12)$$

The presence of several sections with different functions of the area of intersection complicates the process of handling such functions. Therefore, calculation of the function of the area of intersection of a regular $2n$-setsquare in the book is not considered.

We define the function of the area of intersection of simple figures. The graphic description of the process of obtaining the FAI(x) of circle is shown on Fig. 2.6. The FAI(y) of the circle is similarly determined. The functions of the area of intersection of the circle for orthogonal directions of the shift are reflected in the formulas (2.15) and (2.16).

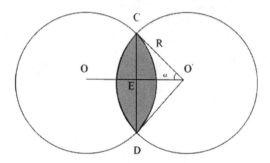

Fig. 2.6: Graphic description of the process obtaining FAI (x) of circle

The distance of the shift $x = OO' = q_x R$, where $q_x \in [0,2]$.

q_x is an auxiliary factor reflecting the relationship of the shift distance and the circle radius.

If we exclude from the formula the radius $R = \sqrt{\dfrac{S_0}{\pi}}$, which is not among the basic elements of the FAI, then $q_x = x \cdot \sqrt{\dfrac{\pi}{S_0}}$

$$\alpha = \arccos\left(\frac{q_x \cdot R}{2 \cdot R}\right) = \arccos\left(\frac{q_x}{2}\right) = \arccos\left(\frac{x}{2}\sqrt{\frac{\pi}{S_0}}\right) \qquad (2.13)$$

$$\alpha = \text{arctg}\left(\frac{2 \cdot CE}{q_x \cdot R}\right) = \text{arctg}\left(\frac{\sqrt{4-q_x^2}}{q_x}\right) = \text{arctg}\left(\sqrt{\frac{4S_0}{\pi \cdot x^2}-1}\right) \qquad (2.14)$$

$$FAI(x) = 2 \cdot \frac{S_0}{\pi} \left(arctg \left(\sqrt{\frac{4S_0}{\pi \cdot x^2} - 1} \right) \right) - x \cdot \sqrt{\frac{S_0}{\pi} - \frac{x^2}{4}} \qquad (2.15)$$

$$FAI(y) = 2 \cdot \frac{S_0}{\pi} \left(arctg \left(\sqrt{\frac{4S_0}{\pi \cdot y^2} - 1} \right) \right) - y \cdot \sqrt{\frac{S_0}{\pi} - \frac{y^2}{4}} \qquad (2.16)$$

If $x = 0$ ($y = 0$), then $FAI(x) = S_0 (FAI(y) = S_0)$.

When identifying the input image as a circle, determine its radius using the basic parameters of the function of the area of intersection.

$$R = \sqrt{\frac{S_0}{\pi}} = \frac{X_{max}}{2} \qquad (2.17)$$

We define the FAI of the rectangle with sides a and b and the tilt angle α for direction shift φ (Fig. 2.7.):

$$A = a \cdot \cos \alpha + b \cdot \sin \alpha$$

$$B = a \cdot \sin \alpha + b. \cos \alpha$$

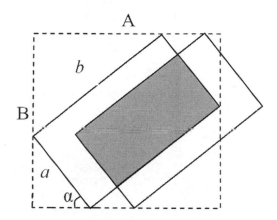

Fig. 2.7: Graphic description of the process obtaining
FAI (x) of rectangle

$$FAI(x) = (b - x \cdot \sin \alpha)(a - x \cdot \cos \alpha) = S_0 - Bx + \frac{x^2}{2} \sin(2\alpha) \qquad (2.18)$$

$$FAI(y) = (b - y. \cos \alpha)(a - y . \sin \alpha) = S_0 - Ay + \frac{y^2}{2} \sin(2\alpha) \qquad (2.19)$$

where

$$A = \frac{S_0}{Y_{max}} + \frac{Y_{max}}{2} \cdot \sin(2\alpha)$$

$$B = \frac{S_0}{X_{max}} + \frac{X_{max}}{2} \cdot \sin(2\alpha)$$

The parameters of the bound rectangles (*A* and *B*) are used only for ease in mapping the process of obtaining the formula of the function of the area of intersection and later replaced by a function of the main parameters (S_0, X_{max}, Y_{max}).

$$FAI(x) = (X_{max} - x)\left(\frac{S_0}{X_{max}} - \frac{x}{2}\sin(2\alpha) \right) \tag{2.20}$$

$$FAI(y) = (Y_{max} - y)\left(\frac{S_0}{Y_{max}} - \frac{y}{2}\sin(2\alpha) \right) \tag{2.21}$$

Formulas (2.17) ÷ (2.20) use the values of the angle of inclination α in the direction of the shift φ. Variants of the formulas of calculating this angle will be described below.

When identifying the incoming image as a rectangle, it is possible to determine the quantitative characteristics of the figure using the basic parameters of the function of the area of intersection.

$$a = \sqrt{\frac{S_0 \cdot Y_{max}}{X_{max}}} \; ; \; b = \sqrt{\frac{S_0 \cdot X_{max}}{Y_{max}}} \tag{2.22}$$

The function of the area of intersection of a square where $X_{max} = Y_{max}$:

$$FAI(x) = (X_{max} - x) \cdot \frac{X_{max} \cdot S_0 - x \cdot \sqrt{S_0 \cdot (X_{max}^2 - S_0)}}{X_{max}^2} \tag{2.23}$$

$$FAI(y) = (X_{max} - y) \cdot \frac{X_{max} \cdot S_0 - y \cdot \sqrt{S_0 \cdot (X_{max}^2 - S_0)}}{X_{max}^2} \tag{2.24}$$

We use principle of triangle similarity for a parallel shift to determine the function of the area of intersection of the triangle:

$$FAI(x) = \frac{S_0 \cdot (X_{max} - x)^2}{X_{max}^2} \tag{2.25}$$

$$FAI(y) = \frac{S_0 \cdot (Y_{max} - y)^2}{Y_{max}^2} \tag{2.26}$$

When we identify the input image of a triangle, we cannot immediately determine the quantitative characteristics of the figure, using the basic parameters of the function of the area of intersection for a pair of orthogonal shifts. This happens because there are four unknowns (three sides of the triangle and the inclination angle of one side to the horizontal). The parameters of the triangle can be determined by its rotation. It is necessary to achieve a situation where the inclination angle of one of the sides to the horizontal is zero. In this case, a large side of the triangle will be equal to the maximum horizontal displacement.

Displacement will be maximum as far as possible. Then, with two sets of basic parameters of the function of the area of intersection, we can determine the quantitative characteristics of the triangle.

A graphical representation of the process of defining one of the triangle parameters sets in Fig. 2.8 is shown.

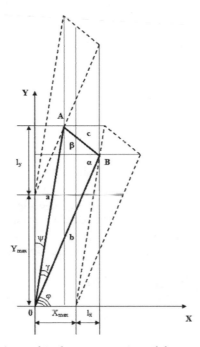

Fig. 2.8: A graphical representation of the process of determining of the triangle parameters

The relationship of basic parameters of the triangle and its quantitative parameters, refer to the following set of formulas.

$$S_0 = \frac{a \cdot X_{max} \cdot \cos \psi}{2} = \frac{b \cdot Y_{max} \cdot \cos \varphi}{2} = \frac{ab \sin \gamma}{2} \qquad (2.27)$$

$$\frac{4S_0^2}{X_{max} Y_{max}} = X_{max} Y_{max} + 2S_0 = ab \cos \varphi \cos \psi \qquad (2.28)$$

$$\frac{a}{b} = \frac{Y_{max}\cos\varphi}{X_{max}\cos\psi}; \frac{X_{max}}{\sin\gamma} = \frac{b}{\cos\psi}; \frac{Y_{max}}{\sin\gamma} = \frac{a}{\cos\varphi} \qquad (2.29)$$

$$b \cdot \sin\varphi = l_x \, ctg\,\psi \qquad (2.30)$$

$$Y_{max} + l_y = a \cdot \cos\psi \qquad (2.31)$$

$$X_{max} + l_x = b \cdot \cos\varphi \qquad (2.32)$$

$$a \cdot \sin\psi = l_y \, ctg\,\varphi \qquad (2.33)$$

$$Y_{max} l_x = X_{max} l_y \qquad (2.34)$$

$$\frac{l_x}{X_{max}} = \frac{l_y}{Y_{max}} = \frac{X_{max}}{b\cos\varphi} = \frac{Y_{max}}{a\cos\psi} = \frac{2S_0 - X_{max}Y_{max}}{X_{max}Y_{max}}$$

$$= \frac{b\cos\varphi - X_{max}}{X_{max}} = \frac{a\cos\psi - Y_{max}}{Y_{max}} \qquad (2.35)$$

$$\frac{X_{max}Y_{max}}{2S_0} = 1 - tg\,\varphi\,tg\,\psi \qquad (2.36)$$

To determine the quantitative parameters of the triangle, we need to perform its rotation. Thus, for example, the side a is to be positioned vertically. Then $a = Y_{max}$, $\psi = 0$, and all the above formulas are much simplified. Sizes of the triangle sides during rotation will not change. In stock will be two sets of formulas. This makes it possible to determine the values of the triangle's sides and the original inclination angle to the horizontal φ.

The function of the area of intersection of the rectangle depends on the inclination angle α to the direction of the shift φ. Depending on the ratio of the sizes of the rectangle (a and b), there are three possibilities of inclination α angle:

$$\sin\alpha \cdot \cos\alpha = \frac{S_0}{X_{max} \cdot Y_{max}} \qquad (2.37)$$

$$tg\,\alpha = \frac{X_{max}}{Y_{max}} \qquad (2.38)$$

$$\sin^2\alpha = \frac{S_0}{X_{max} \cdot Y_{max}} \quad \text{or} \quad \cos^2\alpha = \frac{S_0}{X_{max} \cdot Y_{max}} \qquad (2.39)$$

Options $\sin^2\alpha$ and $\cos^2\alpha$ are identical in further calculations of $\sin 2\alpha$.

On the basis of these formulas, we define the value of sin 2α (the component of the FAI for a rectangle).

$$\sin 2\alpha = \frac{2S_0}{X_{max} \cdot Y_{max}} \tag{2.40}$$

$$\sin 2\alpha = \frac{2X_{max} \cdot Y_{max}}{X^2_{max} + Y^2_{max}} \tag{2.41}$$

$$\sin 2\alpha = \frac{2\sqrt{S_0 \cdot (X_{max} \cdot Y_{max} - S_0)}}{X_{max} \cdot Y_{max}} \tag{2.42}$$

Since the formulas of the intersection area of simple figures is different from each other, the function of the area of intersection can be used for detection of the characteristic features. FAI templates of simple figures are built on the basic parameters (initial area and the maximum shifts).

The basic recognition algorithm of simple shapes does not take into account the effect of noise on the classification process (Aho et al., 1983). It will be as follows:

- Obtaining the real function of the area of intersection in two directions
- Identification of the main parameters of the FAI (S_0, X_{max}, Y_{max})
- Substitution in templates of functions of the area of intersection of the basic parameters that are analytically calculated
- Comparison of actual results of the function of the area of intersection with the analytical calculation gives the function of the area of intersection

The class of simple figures is only a tiny part of the whole class of planar images. It is not possible analytically to calculate the formula of intersection area for most images of varied shapes. In this case, it is necessary to compare the actual results FAI with stored functions for each etalons and for each shift direction. These functions must be obtained at the stage of recognition training.

2.3 Additional Parameters, which Accelerate Plane Image Recognition Process

Image recognition using the function of the area of intersection requires comparison of the actual FAI with an array of etalons FAI that with a large number of etalons requires a lot of time to complete this. It becomes necessary to accelerate the process to introduce certain indicators that would limit the number of etalons FAI that are required for comparison. The function of the area of intersection is continuous and differentiable on the interval from 0 to the maximum shift.

We introduce in the sets of etalons some integral coefficient k_φ, which tends not to change when scaling the basic parameters of the function of the area of intersection (S_0, X_{max}, Y_{max}).

$$k_\varphi = \frac{\displaystyle\int_0^{I_{max}} FAI(i)di}{S_0 \cdot I_{max}} = const \tag{2.43}$$

where φ – the direction for which FAI is determined; i, and I_{max} – respectively are the maximum shift and the shift in the direction φ (Fig. 2.9).

Integral coefficient is determined similarly for $FAI(x)$ and $FAI(y)$ with i replaced by x or y and I_{max} at X_{max} or Y_{max}, respectively. Integral coefficient (IC) for the figures, and the function of the area of intersection of these

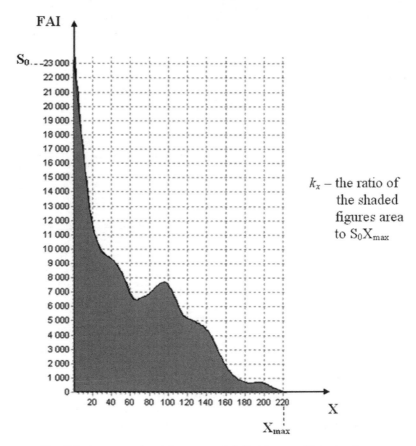

k_x – the ratio of the shaded figures area to S_0X_{max}

Fig. 2.9: An example of the integral definition coefficient using $FAI(x)$ of the image

figures is determined analytically, and can also be calculated analytically.
We calculate the integral coefficient for the circle:

$$\int FAI(x)\,dx = \frac{2 \cdot S_0}{\pi} \cdot \left[\frac{x}{\pi} \cdot \arccos\left(\frac{x}{2}\sqrt{\frac{\pi}{S_0}}\right) - 2 \cdot \sqrt{\frac{S_0}{\pi}} \cdot \sqrt{1 - \frac{\pi \cdot x^2}{4 \cdot S_0}} \right.$$

$$\left. + 2 \cdot \frac{S_0}{3 \cdot \pi^{\frac{3}{2}}} \cdot \sqrt{4 \cdot S_0 - \pi \cdot x^2} - \frac{x^2}{6 \cdot \sqrt{\pi}} \cdot \sqrt{4 \cdot S_0 - \pi \cdot x^2} \right] \tag{2.44}$$

$$k_x = \frac{\displaystyle\int_0^{X_{max}} FAI(x)\,dx}{S_0 \cdot X_{max}} = \frac{4}{3 \cdot \pi} = const \tag{2.45}$$

We calculate the integral coefficient for the rectangle.
The general formula IC of the rectangle will be as follows:

$$k_x = \frac{1}{2} - \frac{1}{12} \cdot \frac{X_{max}^2}{S_0} \cdot \sin 2\alpha \tag{2.46}$$

Considering three options of possible values of the angle 2α, we get three options for the formulas of integral coefficient for the rectangle.

$$k_{x_1} = \frac{1}{2} - \frac{1}{6} \cdot \frac{X_{max}}{Y_{max}} \tag{2.47}$$

$$k_{x_2} = \frac{1}{2} - \frac{1}{6} \cdot \frac{X_{max}^3 \cdot Y_{max}}{S_0 \cdot \left(X_{max}^2 + Y_{max}^2\right)} \tag{2.48}$$

$$k_{x_3} = \frac{1}{2} - \frac{1}{6} \cdot \frac{X_{max}}{Y_{max}} \sqrt{\frac{X_{max} \cdot Y_{max}}{S_0} - 1} \tag{2.49}$$

We calculate the integral coefficient for the square, where $X_{max} = Y_{max}$.

$$k_x = \frac{1}{2} - \frac{1}{6} \sqrt{\frac{X_{max}^2}{S_0} - 1} \tag{2.50}$$

Integral coefficient for the triangle will be as following:

$$k_x = \frac{1}{3} = const \tag{2.51}$$

Formula integral coefficients of simple figures are shown in Table 2.1.

Table 2.1: Formulas of the integral coefficients of simple figures

Name of the simple figure	Integral coefficient
Circle	$\dfrac{4}{3\cdot\pi}$
Triangle	$\dfrac{1}{3}$
Square	$\dfrac{1}{2}-\dfrac{1}{6}\sqrt{\dfrac{X_{max}^{2}}{S_0}-1}$
Rectangle	$\dfrac{1}{2}-\dfrac{1}{12}\cdot\dfrac{X_{max}^{2}}{S_0}\cdot\sin 2\alpha$

For images that are not simple figures can also obtain the value of the integral coefficients (Belan and Yuzhakov, 2013; Koh et al., 1994; Vetter and Poggio, 1997). For this, we must calculate the area of real FAI and divide the value by the product of the initial area and the maximum shift. These coefficients do not change under affine transformations.

Integral coefficients are appropriate to apply to the image recognition algorithm in the comparison step. Before the phase of the comparison of FAI, it is necessary to carry out a comparison of the IC. Application of integral coefficients will substantially reduce the number of samples for which comparison of functions are performed.

The coincidence of the IC in images is a necessary condition but it is not sufficient for classification. Whatever additional parameters are used to speed up the recognition process, only the values of the functions of the area of intersection carry information about the shape of the object.

Also integral coefficients of the image in the processing should be used with other factors that may accelerate the process of classification of images and determining their parameters.

$k\rho_{\varphi}$ is the coefficient of elements density of the image to the direction φ.

$$k\rho_{\varphi}=\frac{S_0}{X_{max}\cdot Y_{max}} \tag{2.52}$$

This coefficient reflects the degree of contour images. In general, it can vary from 0 to 1, for example, for a square ($X_{max}=Y_{max}$) according to the spatial orientation. $k\rho_{\varphi}\in\left[\dfrac{1}{2};1\right]$. Only for a rectangle with sides a

and b (where $2a < b$) and the inclination angle of one of the parties to the direction of the shift $\alpha = 45°$ can this coefficient take a value $k\rho_\varphi = \dfrac{b}{2a} > 1$. This coefficient of arrangement density can be used for classification of the rectangles with the specified parameters.

kE_φ is a coefficient of the object eccentricity to the direct φ.

$$kE_\varphi = \frac{Y_{max}}{X_{max}} \tag{2.53}$$

This coefficient can be used for scaling the process of determining the image difference and its projection on the plane of the receptor field that is needed in the algorithm for determining the spatial orientation of the object. For figures with identical shapes, this parameter in the chosen direction is same.

A parametric coefficient kP can be used to speed up the recognition process. It is an area of the image that is created by the integral coefficients, which are presented in the form of a parametric function.

$$kP = S(k_\varphi) \tag{2.54}$$

kP coefficient is useful if there is a very large number of etalons and only when the comparison of integral coefficients of etalon with actual values occupies a large amount of time. However, this factor requires very high accuracy calculations. As the value of the integral coefficients belongs to the range from zero to 0.5, it can have a value of not more than $0{,}25\pi$.

Recognition stage, which uses a variety of additional coefficients, is called the fast recognition stage. Recognition stage, which compares the actual and the pattern function of the area of intersection is called the detail recognition stage. This stage compares the functions. However, its duration is higher than in the fast stages. Overall image recognition algorithm is presented below.

The coefficients $kP, k_\varphi, k\rho_\varphi, kE_\varphi$ are invariant with respect to the image size and its location on the receptor field. They are calculated on the basis of the basic parameters of the function of the area of intersection, and, therefore, their use is not contrary to the concept of parallel shift technology for image classification.

2.4 Algorithm of Recognition Images Based on PST

The functions of the area of intersection obtained at the pretreatment step should be specifically analysed in order to identify the image. The function of the area of intersection and different coefficients previously discussed can be used during recognition as the method to compare with etalons for neural networks.

In the case of using neural networks, the function of the area of intersection and their coefficients can serve as a vector for the input of the characteristic features of artificial neurons. This detection method does not require a large amount of expenditure for information storage.

This book is considered by means of image recognition method for reconciliation with etalons. It can be divided into two blocks—the first is to identify simple images, which FAI can be determined analytically; the second one is recognition of all other figures.

Before performing, the image recognition process is subjected to a pretreatment. The result of this process is a binary image. For this method, the image in the receptor field will be considered holistic, even if it consists of several individual parts. Image preprocessing using parallel shift technology will be discussed in the following sections.

The basic algorithm of recognition of simple figures has already been described. In reality there are many factors that can lead to loss of information. These include, for example, loss of information during rounding off calculations and when digitising images or noise impact. Therefore, it is necessary to use values of certain permissible deviations in the process of comparing.

The result of the first recognition unit may be a decision on the identification of the image as a simple figure and getting its qualitative and quantitative parameters. The advantage of its selection from the general algorithm of recognition is the fact that in the case of identification of the object at this stage, you do not need to search among an array of etalons. This will significantly reduce the recognition process. Furthermore, the cost of storing information for this block is minimal. Etalons of simple figures are stored as templates of analytical formulas, which will be filled with the basic parameters of the function of the area of intersection.

If in the previous stage a decision was not made on the identification of the image as a simple figure, it is necessary to switch to the recognition by a comparison of figure with an array of etalons. Each of the recognition units is in turn divided into two steps—the first stage is the fast recognition, which compares the integral coefficients; the second stage is detailed recognition where the function of the area of intersection is compared. Scheme of recognition process is shown in Fig. 2.10.

Overall image recognition algorithm based on the analysis FAI consists of the following steps:

1. For obtaining real images the function of the area of intersection in two directions (φ and $\varphi_1 \neq \varphi$).
2. Identification of the main parameters of these (S_0, X_{max}, Y_{max}) and real-IC (k_x and k_y).
3. In substitution IC in the formula of simple figures, values of key parameters are derived from the function of the area of intersection.

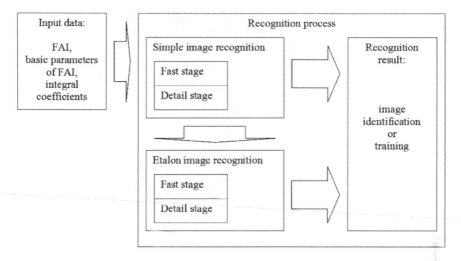

Fig. 2.10: Scheme of recognition process

Comparison with specified accuracy dk_1 is one of two real-IC (k_x) with analytically calculated IC of simple figures. We get the class XS figures, that have matches.

4. Comparison with specified accuracy dk_1 of other real IC (k_y) to calculate IC for class XS figures. We get YS class of figures, that have matches $(YS \in XS)$.

5. Comparison with specified accuracy dS_1 of real FAI (y) with analytically calculated FAI YS-class of simple figures. We get the class of figures $YS1$, that have matches $(YS1 \in YS)$.

6. Comparison with specified accuracy dS_1 real FAI (x) with analytically calculated FAI $YS1$-class of simple figures. We get the class of figures $XS1$, that have matches $(XS1 \in YS1)$.

7. When properly chosen values of deviations dk_1 and dS_1, are taken, then the image can be identified as a specific figure.

8. If at any stage of the action of the algorithm produces an empty class of figures $(XS = \varnothing, YS = \varnothing, YS1 = \varnothing, XS1 = \varnothing)$, then the image is not classified as simple. We must move to the reference image recognition unit.

9. If the number of figures class $XS1 > 1$, then it is necessary to adjust the value dk_1 and dS_1 and repeat steps $3 \div 6$ of the algorithm.

10. A comparison with a given accuracy dk_2 of one of the two real IC (k_x) with IC of all etalons database objects is done. We get the class of figures X, that have matches. The φ direction of class X figures that have matches is recorded.

11. Comparison with specified accuracy dk_2 of other real IC (k_y) with IC class of figures X etalons database in directions $\varphi 1$. We get the figures class Y, that have matches $(Y \in X)$.

12. Comparison with specified accuracy dS_2 of real FAI (y) with FAI of class of figures Y reference database in directions $\varphi 1$. We get the class of figures $Y1$, that have matches $(Y1 \in Y)$. In the process of comparison, it is necessary to scale the etalons FAI for bringing their parameters to the parameters of the real FAI (y).

13. Comparison with specified accuracy dS_2 of real FAI (x) with FAI of figures $Y1$-class etalons database in φ directions. We get the class of figures $X1$, that have matches $(X1 \in Y1)$. In the process of comparison, it is necessary to scale the etalons FAI for bringing their parameters to the parameters of real FAI (x).

14. When properly chosen values of tolerances dk_2 and dS_2, then there is a corresponding image among an array of input image clearly recognised standards (number of figures class $X1 = 1$).

15. If at any stage of the action algorithm receives an empty class of figures $(X = \varnothing, Y = \varnothing, Y1 = \varnothing, X1 = \varnothing)$, there is no etalons for a given image or no correctly set tolerances dk_2 and dS_2. It is necessary to adjust the value dk_2 and dS_2 repeat steps $10 \div 13$ of algorithm or conduct learning system to create a new etalon.

16. In the number of figures class $X1 > 1$, it is necessary to adjust the value dk_2 and dS_2 and repeat the steps $10 \div 13$ of the algorithm.

The values of the relevant deviations for both detection units (dk_1 and dk_2, dS_1 and dS_2) may be the same or different. Recognition systems can be used as fixed value tolerances and select them individually for each etalon.

Software implementation of the algorithm is an option when $10 \div 13$ steps in the cycle are performed sequentially for each etalon. This simplifies access for software implementation of the algorithm which is an option when $10 \div 13$ steps in the cycle are performed sequentially for each etalons database.

In the input image pixel dimensions $D \times D$, the performance division detection method is increased in two stages at about $2D^2$ versus time by direct comparison of arrays.

The proposed image recognition method is based on criteria like invariant with respect to affine transformation that is a parallel shift. When the figure is resized, its function of the area of intersection does not change form and are easily scalable in size and integral coefficients do not change. No need to keep the definition of the input image of the function of the area of intersection in all directions, as the value of the function of the area of intersection to the etalons is stored at a etalons' surface. As a result of this, image recognition is possible regardless of its angle of rotation with the respective etalon.

2.5 Processing of 3D-Objects through PST

The principles of processing parallel shift technology by using objects with more than two dimensions are similar to the processing of plane images. For the form of 3D-objects, the figure describes a set amount of functions of the volume of intersection.

The function of the volume of intersection (FVI) depends on the volume of intersection of the original object and which is shifted parallel to the selected direction, from a distance of displacement. The shift to the technical implementation of convenience is advantageously carried out in three orthogonal directions. The basic parameters of the function of the volume of intersection will be the initial volume of the object (V_0), and the maximum values of shifts in orthogonal directions (X_{max}, Y_{max}, Z_{max}).

Also, there are simple objects that FVI can be determined analytically using the basic parameters.

So, the function of the volume of intersection for sphere will be as follows.

$$FVI(x) = V_0 - \frac{\pi \cdot X_{max}^2 x}{4} + \frac{\pi \cdot x^3}{12} \tag{2.55}$$

$$FVI(y) = V_0 - \frac{\pi \cdot Y_{max}^2 y}{4} + \frac{\pi \cdot y^3}{12} \tag{2.56}$$

$$FVI(z) = V_0 - \frac{\pi \cdot Z_{max}^2 z}{4} + \frac{\pi \cdot z^3}{12} \tag{2.56}$$

Integral coefficient of the sphere:

$$k_\varphi = \frac{\int_0^{I_{max}} FVI(i)\,di}{V_0 \cdot I_{max}} = \frac{3}{8} = const, \tag{2.58}$$

where φ is direction, which is determined for the function of the volume of intersection; i and I_{max} are respectively the maximum shift and the shift in shear direction is φ.

While identifying the object as a sphere, we can calculate the radius:

$$R = \sqrt[3]{\frac{3V_0}{4\pi}} = \frac{X_{max}}{2}. \tag{2.59}$$

To determine the function of the volume of intersection of the triangular pyramid we use similarity property figures. Then the function of the volume of intersection will be as follows.

$$FVI(x) = \frac{V_0(X_{max} - x)^3}{X_{max}^3} \tag{2.60}$$

$$FVI(y) = \frac{V_0(Y_{max} - y)^3}{Y_{max}^3} \qquad (2.61)$$

$$FVI(z) = \frac{V_0(Z_{max} - z)^3}{Z_{max}^3} \qquad (2.62)$$

Integral coefficient of the triangular pyramid is:

$$k_\varphi = \frac{\displaystyle\int_0^{I_{max}} FVI(i)\,di}{V_0 \cdot I_{max}} = \frac{1}{4} = const, \qquad (2.63)$$

For the cube and, moreover, for arbitrary parallelepiped to calculate the function of the volume of intersection for their basic parameters the FVI is quite difficult. Also unknown lengths of the sides of the arbitrary parallelepiped will be the angles of inclination to the orthogonal planes. Object data processing can be performed as with the arbitrary shape of figures.

Recognition algorithm will be the same as for plane figures. Just are added a paragraph to compare the integral coefficients and the function of the volume of intersection that are obtained for z direction.

The set of functions of the volume of intersection for each etalon will form a certain hypersurface. This hypersurface will have central symmetry; so it must be stored for directions and forming a hemisphere.

The possibility of manipulation on the basis of parallel shift technology with 3D-objects is considered purely theoretically. At present there are no effective detectors of the object volume and volume of the intersection of two objects.

3

Methods of Pre-Processing Images Based on PST

Analysis of the video information cannot be without its pre-treatment. There are currently a large number of tools and methods for the process. In different sources, they are called in different ways and the order of these actions is not uniquely determined.

In general, pre-treatment of images may be divided into two components—modification and transformation.

Modification involves performing of certain actions on the object with a view to change the input image to another; one that is more suitable for further processing. Modifying process includes, for example, pre-processing, image enhancement, binarisation and noise removal. The result of this process is production of a corrected image, which is slightly morphologically different from the input image.

Transformation is a sequence of actions that give the result in the receipt of a set of image properties allowing further analysis of this or that method. The process of transformation includes clipping, division into layers, edge detection and selection of skeleton, segmentation. The result of this process is the corrected image in a certain way that is significantly different from the original, or a set of specific properties of the original image.

Pre-treatment process of the video information is separated into two components and is correlated with the classification of the raster image processing methods that is described in Rafael C. Gonzalez papers and Richard E. Woods (Gonzalez et al., 2004). They share these methods into three classes—low, medium and high. Methods of low level are applied to input and output images; mid-level methods to the image input and on the output with a set of properties and attributes; and the high-level methods provide analysis of the input digital image to produce certain qualitative and quantitative characteristics of the object.

Thus, modification of the image can be attributed to methods of low pretreatment. The process of transformation cannot unambiguously hold a similar definition. Such converting steps, as object selection and the edge detection methods, are obviously is low-level methods. Other transformation methods are used depending on the methods for further analysis and have a set of image properties on the output. Therefore, they belong to the image-processing methods of mid-level.

Existing pre-treatment video information methods include the use of rather complex mathematical calculations, especially concerning the various filtering methods (frequency filtering, noise filters). Effectiveness of their use is based on the increasing power of computer technique.

Modern signal preprocessing means include its conversion to digital form, because of which are called digital signal processing (DSP). In the general case, DSP complex consists of a computer, dedicated hardware and software. Personal computers may be used for solving problems with low demands on information processing speed as the analysis unit.

Preprocessing of images based on parallel shift technology requires the creation of specialised means. At this stage, the process carried out by the authors uses the software model.

Use of parallel shift technology for image processing is convenient to demonstrate on the example of Chinese characters. They have complex and varied shapes, with their shape being important and thus are suitable for creating binary images. In this book, a variety of image processing methods are used by the Chinese character for 'dragon' (Fig. 3.1).

To create the possibility of applying image processing methods based on a parallel shift technology, the following must be done.

Fig. 3.1: Image of Chinese character 'dragon'

- Delamination is a separation of the image into several grades according to the brightness. The number of luminance ranges is selected depending on the need to address those or other problems.
- Binarisation is transformation of the image into a binary matrix. This process is based on a comparison of the brightness of each pixel with the luminance threshold value. The value of the brightness of the pixel above the threshold of brightness in the binary image corresponding to a pixel is set to 0, and otherwise to 1.
- Denoise is reduction existing in image extraneous information, which is not part of the input image and appears due to certain actions of external factors. A typical method of execution of the process is to remove the high frequency components of the signal spectrum.

- The next stage of preliminary image processing is the process of obtaining a set of functions of the area of intersection.

The processes of the bundle and image binarisation are simple enough. Their implementation has no direct relation to the description of the parallel shift technology. Therefore, these processes in this book will not be considered in detail. These steps will be most important stages in analysis or raster image processing. This chapter will focus more on the process of removing noise.

3.1 Edge Detection Method of Plane Figures

Description of image preprocessing based on the technology of parallel shift will start by describing the process of edge detection of the binary image. This process was not declared previously as part of preprocessing. However, shown below is that edge detection has one of the stages of removal process in the binary image noise using parallel shift technology.

The edge detection of image is itself one of the main tasks of the video information processing systems (Petrou and Kittler, 1991; Walsh and Raftery, 2002; Belan, 2011; Spacek, 1986). Although PST involves manipulation preferably without edge objects, its use will allow detection of edge of the image on the stage of pretreatment.

Here we would like to stop on the concept of object that is presented as an edge. This term in relation to images from different sources is treated differently. The contour printing image makes a clear distinction between the subject and the background. It is also possible to interpret the contour image as a barcode or image consisting of thin lines. The second interpretation will be used in this book, for example, the papilla-lines figure has an image with a high edge value.

At the moment, there are many methods for edge detection of the image, which are used in almost in all the existing image processing and recognition systems. This is due to the fact that the edge carries most of the image information of the object. During digital data processing, edge detection occurs through the use of various methods of masking. One of the main disadvantages of their application is the need to replace the mask when the edge width is changed. Furthermore, with increasing width of the initial edge, the size of the mask grows too. Therefore, the number of operations and time required for edge detection increase.

One of the advantages of edge detection based on parallel shift technology is the possibility of edge detection of a binary image of any predetermined width. The proposed method for edge detection can be used both for digital and non-digital image processing. To implement this process it is necessary to perform a small number of operations.

The shape that the edge will assume is a layer of given width h, which belongs to the figure and is on the border with the background. This value may be expressed in pixels in digital image processing. With non-digital image processing the h value can have a distant value.

Let us consider the edge detection procedure for the variant of processing the digital information. In this case, the edge width h can be expressed in pixels. All the elements of the image edge of desired value of width h belong to the set C_{rez}. All elements of the image edge belong to the set of C_i, if the copy on the i units is being shifted.

Then

$$C_{rez} = \sum_{i=1}^{h} C_i.$$ (3.1)

It should be noted that $C_{rez} \neq C_h$.

The C_i sets definition can be realised in two ways. For obtaining FAI, the shift of copy can be performed cyclically, as will be discussed in the following chapters. The second method is shift of the image copy in the selected directions by a predetermined number of pixels (or a predetermined distance in non-digital image processing). This method is much simpler to implement in hardware.

Machine vision systems use the Euclidean coordinate system for the receptive field in digitalisation and shifts easily to organise in orthogonal directions and diagonally. Therefore, during software simulation of edge detection process, the shifts to eight directions were used.

Part of the process of image edge detection obtained for $h = 3$ pixels is shown in Fig. 3.2. In this case, for edge points, the shifts of the copy figure in the four orthogonal directions are shown.

a) initial image   b) right shift   c) left shift

d) down shift   e) up shift   f) contour

Fig. 3.2: Process of edge detection with $h = 3$ pixels

Black on all the figures marks the points belonging to the original image while everything else are the points of image copy and intersection area. Figures present shifts of copies by 3 pixels in orthogonal directions. The amount of black elements on them gives a set of the edge elements C_3. Images of shifts for localised elements of C_1 and C_2 were not shown because these images do not differ from those offered. Therefore, they aren't informative and give little understanding of the edge detection process. Shifts to the diagonal directions are not shown for the same reasons.

However, during edge detection with h width, all C_i are always present for orthogonal and diagonal shifts.

The technical realisation of the process of edge detection based on the technology of parallel shift may lead to certain errors. Let it be determined that the edge of specified width in the generated image is as shown on Fig. 3.3. It should be noted that despite the fact that the picture consists of two objects, it is a considered as one.

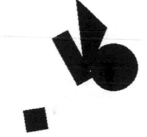

Fig. 3.3: Image for which the contour is determined

When the edge detection with a width of 20 pixels is done by shifting the copies to eight directions, the following results are obtained (Fig. 3.4).

Errors of the process in this case will be incomplete determination of the edge (wrong side) and incomplete edge detection around the individual empty pixels (snowflakes). These errors occur when there is the image of interior angles (wrong side) and small voids (snowflakes).

Problem-solving methods in the process of edge detection are the next:

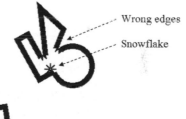

Fig. 3.4: Contour definition errors when $h = 20$

1. *Preprocessing of the image:* In the pretreatment process, the presence of small voids is eliminated from the binary image.
2. *Increasing the number of shift directions:* This will provide a more detailed contour and increase in the number of branches of the snowflakes, bringing it closer to the real contour of the desired width.
3. *Reducing the width of the desired edge:* When the desired width of the edge is less than four, the mentioned error is not visually noticeable.
4. *Investigation of primarily convex objects:* This eliminates the cause of the internal angles. The main advantage of the proposed method is the ability to detect image edge of any given width.

3.2 Use of Parallel Shift Technology for Noise Removal in Images

In the real world, information perception by any intelligent system is accompanied by the presence in a signal of a certain amount of noise among useful data. Misinformation arises owing to the effects of certain internal and external factors.

The intelligent system cannot influence the external factors without changing the environment. Reduction in the amount of noise can build new means of signal detection.

Internal factors of negative impact are lost during transmission of information within the system and the error in the chosen method of noise control. In processing the signals, one of the methods of noise control is frequency filtering. Digital image processing is widely used by a variety of masking. It allows smoothening out of sudden changes in pixel brightness of the object. Also, masking allows selection of certain elements of the image for further removal from the visual field.

The use of parallel shift technology allows separation of the noise removal process into three stages. Since use of the parallel shift technology analysed the shape of the image, the object of study will be considered as a binary image. Half-tone images are divided into binary image at the stage of the bundle. Assume that noise information is evenly distributed across all levels of brightness of the image. Then, if the image is divided into brightness gradations of ranges and converted into n binary objects, in each range hit only a $\dfrac{1}{n}$ part of all non-zero elements of noise. Noise elements with zero value will be duplicated in each layer. Expected number of noise information that hits into each binary object is less than the noise amount of the original image. Thus, the separation process can be considered as the first step in removal of noise.

The second stage of the process should be removal of noise by using parallel shift technology. As we study the binary image, the noise in this case can be considered a stand-alone group of single elements and zero elements that distort the object. In the process of software simulation following types of noise were used:

1. *'Pepper' type of noise:* Equal filling of the visual field of a binary image by cells that have '1' value.
2. *'Salt' type of noise:* It is equal filling of the visual field of a binary image by zeros.
3. *'Salt' and 'pepper' type of noise:* Here it is a combination of the two previous types.

In the noise removal process, use of parallel shift technology is necessary to apply in the edge detection method described above. In this case, all parts of the image, defined as contour images, must be removed. The larger the width of the desired contour *h*, the greater the size of the objects to be considered as noise and removed from the visual scene. But in this case, much of the useful information about the image would be lost. A large number of items that belong to the wide border of the object will disappear. Significant increase in noise occurs in the inner regions which are formed by the second type of noise. To minimise the impact of the removal noise process on the image contour, we defines the minimum width *h* = 1. Examples of noisy image character 'dragon' and results of the removal of the edge points at level 34 dB of noise are shown in Fig. 3.5.

It is necessary to assess the loss of information when removing the edge points and the ability to identify an object in this case. In Fig. 3.6 is given FAI (*x*) of the character 'dragon' without noise inclusions and in Fig. 3.5 is seen the presence of noise such as 'pepper'. In the presence of noise such as 'salt' and 'pepper', PSNR = 34 dB.

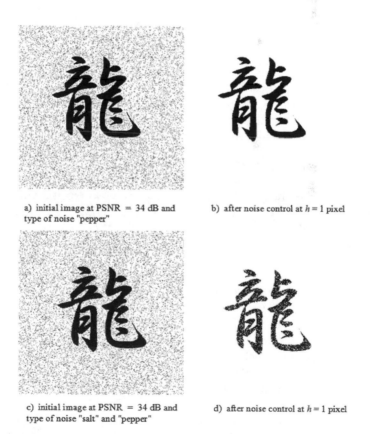

a) initial image at PSNR = 34 dB and type of noise "pepper"

b) after noise control at *h* = 1 pixel

c) initial image at PSNR = 34 dB and type of noise "salt" and "pepper"

d) after noise control at *h* = 1 pixel

Fig. 3.5: Use of the contour identifying process for noise control

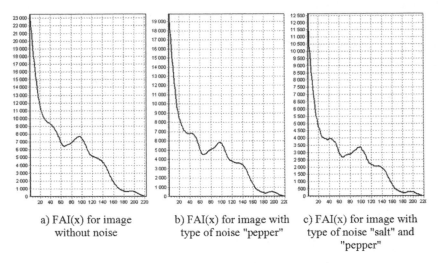

| a) FAI(x) for image without noise | b) FAI(x) for image with type of noise "pepper" | c) FAI(x) for image with type of noise "salt" and "pepper" |

Fig. 3.6: FAI(*x*) of image for different types of noise

In the result of removing the points of the image edge, the values of the function of the area of intersection for images, in which there are various types of noise, are different. However, it should be noted that, in general, the FAI form remains the same. Therefore, when comparing the functional area of image intersection after noise removal with the same functions of the pattern images, it's necessary to consider scaling factor k_{sc}.

$$k_{sc} = \frac{S_{0real}}{S_{0etalon}} \tag{3.2}$$

where S_{0real} is the real image area after removal of the contour points, $S_{0etalon}$ is the area of the pattern image. In addition, when comparing the image FAI, it's necessary to remember that the image sizes may differ, too. Pictures shown on Fig. 3.5 are of the same size. Therefore, in general scaling is required by x-axis of this function. The concept of the pattern image will be described further. For other directions, that are different from the above example, the direction of *x*, calculations are carried out similarly.

Here is an example of calculating certain coefficients to describe the real image based on the scaling coefficient k_{sc}. To adjust the real coefficients should be multiplied by k_{sc}. Suppose there are three images of character 'dragon' after deleting a different noise level (Fig. 3.7). Noise levels correspond to 30 dB, 26 dB, 20 dB. The process of noise control is done by deleting the edge points.

Values of some real and adjusted coefficients are shown in Table 3.1.

The adjusted ratio values are different from the reference coefficients in no more than 10 per cent. Therefore, in the noise range, where PSNR

a) initial image

b) after noise control

Fig. 3.7: Images of character 'dragon' before and after the process of noise control

Table 3.1: Values of some real and adjusted coefficients

Coefficient or parameter	Without noise	PSNR, dB		
		30	26	20
Initial area, S_0	23554	14374	11642	7830
Integral coefficient in x-direction, k_x	0,23665	0,14999	0,12042	0,08319
Integral coefficient in y-direction, k_y	0,28344	0,18105	0,14766	0,10033
Density coefficient, $k_{\rho\varphi}$	0,44393	0,27685	0,22023	0,15279
Scaling coefficient, k_{sc}	1	1,63865	2,02319	3,00817
Adjusted k_x	0,23665	0,24578	0,24363	0,25024
Adjusted k_y	0,28344	0,29667	0,29874	0,30180
Adjusted $k_{\rho\varphi}$	0,44393	0,45366	0,44556	0,45963

is 40 to 20 dB, removal of the edge points in the fight against noise can be used.

The third stage of removing the noise process is adjustment of the function of the area of intersection at its creation. Here is an example. Suppose that after the removal of the edge points, the image remains as is shown in Fig. 3.8a, and its FAI (*x*) is shown in Fig. 3.8b.

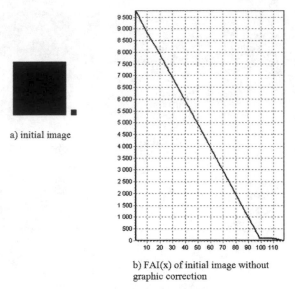

a) initial image

b) FAI(x) of initial image without graphic correction

Fig. 3.8: Example of causing correction in FAI graphs

In this example, the image of square 100 × 100 have the main image and square 10 × 10 is the noise, which is not removed in the process of removal of noise. If the noise is in the direction of the shift copy of the original image, then the formation of function of the area of intersection may cause errors. In a graph FAI (x) of Fig. 3.8b there is a horizontal component, which does not allow determination of the maximum shift correctly. In this case, it becomes necessary to correct the graph.

We introduce the concept of the value of noise effects threshold (dN). DN value specifies a certain percentage of the elements of the input image area (S_0), which will be considered as noise. Thus, the correct formula of the basic parameters of the function of the area of intersection, as maximum shifts, takes the following form:

$$X_{max} = x - 1, \text{ if } FAI(x - 1) = FAI(x) \text{ and } FAI(x) \leq dN \cdot S_0 \tag{3.3}$$

$$Y_{max} = y - 1, \text{ if } FAI(y - 1) = FAI(y) \text{ and } FAI(y) \leq dN \cdot S_0 \tag{3.4}$$

The values obtained from these formulas are fully consistent with options for obtaining the value of the maximum shifts using zero-crossing property.

Note that the functions of the small horizontal sections crossing areas can be only a consequence of the form of the input object. In this case, those elements of the original image will be perceived as noise. Using a threshold value of noise components, we cut the excess information in FAI graphs in the process of construction.

There are other methods of determining the shape of a figure apart from using the functions of the area of intersection (Belan and Motornyuk, 2013); for example, one can simply scan the image to determine the number of elements which are arranged on a line perpendicular to the scanning direction. For example, when scanning a square orthogonally arranged 100×100 functions equal 100 can be obtained in the area of 0 to 100 (Fig. 3.9).

Lets carry out the procedure for comparing the noise immunity PST use in the process of determining the shape of the object to the method of scanning the figure sizes.

Let us assume that after the removal of interference by some method of noise control, together with the square 100×100 image includes square 10×10 of noise components (Fig. 3.8a). Then, while scanning vertically, we obtain a function $x = f(y)$, which is at y values between 0 and 10 that will be equal to 110. This is 10 per cent more than the value of 100 obtained by scanning square 100×100 without noise inclusion. In addition, it is not clear how to interpret the possible presence of empty cell groups between the unit cells of an arbitrary image. This case may occur when there are voids inside the shape, which is analysed.

Carry out the definition FAI (y) of the same square with noise (Fig. 3.8a). The initial image area (S_0) will be equal to 10100, which is 100 units more square area 100×100 without noise inclusions. At the section x from 0 to 10 FAI deviations due to the influence of the noise components is reduced from 100 to zero. Influence of noise when determining FAI (x) is levelled by applying a threshold of the noise elements. The maximum deviation of the functions of the area of intersection in the example is close to 1 per cent. So, noise immunity of the process of determining the shape of the object using a parallel shift in technology is much higher than the scan size figures. This occurs due to the fact that the object, which is determined by deviation PST is the application number of picture elements. When scanning shape sizes of that object, which is defined by

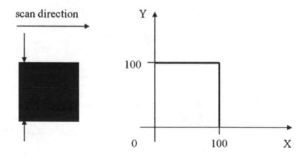

Fig. 3.9: Example for production of object shape function
$y = f(x)$ by scanning it in a horizontal direction

deviation, it is the number of image elements that are located between two of its borders on the line scan.

We can conclude that by increasing the dimension of the object, it is possible to increase the number of elements constituting it. Therefore, it is increase of base number, which is determined by the presence of noise rejection. So, it is increase in the noise immunity of image processing techniques that utilise parallel shift technology.

When comparing the image processing algorithms functions, the value of the permissible error dS can be used. The question arises whether this parameter is able to compensate for the deformation of the figure due to the presence of noise and distortion caused by the noise removal process. Figure 3.10 shows the graphs of standard FAI and image of the function of the area of intersection after removal of noise when noise inclusions percentage (pN) is 10 per cent. The explanation of the parameter pN and its relationship with the PSNR parameter are given in Chapter 6. The real function of the area of intersection by Fig. 3.10 is scaled by using the scaling factor k_{sc} (Formula 3.2). We use the method of removing the contour points to noise control.

The real and template images belong to the same class. For a good comparison of their functions of the area of intersection it is necessary to

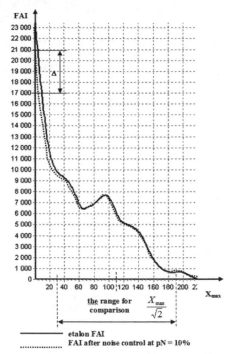

Fig. 3.10: The use of part of the function of the area
of intersection to compare functions

have value dS parameter more than Δ deviation. However, this results in increased permissible error values. So, at a detailed stage of functions, comparison to one class of different images can be assigned. As seen in Fig. 3.8, the maximum distortions of functions of the area of intersection occurs at the beginning or near to the value of maximum shift if there are deformation factors. This depends on the shift direction. In any case, the initial area is changed due to the presence of noise. If, in the direction shift of image copy, there are inclusions of noise, then significant FAI distortions may occur at the end of the function. This is analogous to the factors which lead to the use of the permissible error dN.

It is necessary to limit the increase in the value of the parameter dS. To resolve this issue on a detailed comparison stage, not all values function of the area of intersection are used; only a range of values, where the expected deviation is minimal can be taken into account. Figure 3.10 shows that the length of this range has the value of 70 per cent of the maximum shift in the selected direction.

Since the limitation of range values for functions comparison helps to neutralise the influence of image distortion, it can be considered as the fourth level of noise control.

3.3 Technical Methods for FAI Preparation

The main objective of flat image preprocessing, when using parallel shift technology, is to convert the two-dimensional image into a set of functions of the area of intersection.

As shown previously, the function of the area of intersection is inextricably related to the shape of the object. So their use is possible for the image recognition process. The process of obtaining the functions of image area of intersection and its copies is easy to implement in hardware.

When digital image is processed, the process of obtaining the function of the area of intersection takes place as follows:

1. For the input image array size of $k \times m$, consisting of elements $a_{i,j}$ ($a_{i,j} \in [0;1]$) creates a copy (array of elements $b_{i,j}$).

2. $S_0 = \sum\limits_{i=1}^{k} \sum\limits_{j=1}^{m} a_{i,j}$

3. It starts a parallel shift of image copies in the selected φ direction.

4. At every shift t step intersection array which consists of elements $c_{i,j}$ ($c_{i,j} = 1$ if $a_{i,j} = 1$ and $b_{i,j} = 1$ is created; otherwise it is $c_{i,j} = 0$).

5. $FAI(t) = \sum\limits_{i=1}^{k} \sum\limits_{j=1}^{m} c_{i,j}$

Hardware process can be implemented as follows. We use two matrices (A and B), one of which (B) is a set of shift registers. Depending on the type of functions obtained in the intersection area, the shift may be cyclic or not. Reasons for obtaining various types of the function of the area of intersection will be described in the following chapters. The matrix A is filled with the input image elements $a_{i,j}$. This A matrix is copied into matrix B, in which the shift occurs by a signal of control device. Appropriate cell arrays combine elements of logical multiplication. The outputs of U gates are fed to summation block-level output signal and the value of which corresponds to the FAI value for the current time shift. Construction of the summation block may be implemented using standard logic components (encoders, counters, multiplexers, analog-to-digital converters). Rotation of the functions of area of intersection should be carried out continuously until the stop signal system is reached. Non-cyclic shift is performed before the absence of the intersection shape and its copies, or when reaching the FAI threshold with the influence of noise effects threshold (dN). Block diagram of the device receiving FAI on Fig. 3.11 is shown below.

In the functions of the area of intersection, the flat binary image is considered as image on input. The matrix elements are set to 0 or 1. To study the halftoned image, the gradients of luminance change at appropriate points to be seen as elements of input matrix. This opens up broad prospects for application of different methods of interaction between the elements of the image and its copy. Studies have not been conducted in this direction though.

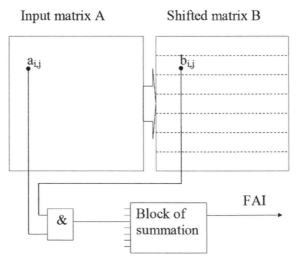

Fig. 3.11: Graphical representation of the block diagram of the device receiving FAI

Obtaining functions of the area of intersection is the main goal of the preliminary imaging process using parallel shift technology. It can be attributed to transforming of information from one form to another.

3.4 Model for Obtaining FAI of No Bitmap

Most of the known pre-treatment methods consider the image as a bitmap. The object transformation method into a set of functions of the area of intersection allows processing of a bitmap image, as well as no bitmap. When transactions with no bitmaps occurs, the preprocessing and binarisation processes are not necessary. The process of separation can be realised by means of optical filters. Possible model for the implementation process of FAI in this case is shown in Fig. 3.12.

Marking time t_1 corresponds to the moment the path of the light flux area of image S_0 appears. Time t_2 is a moment of zero crossing shape and its copy, which is shifted in parallel.

In this case, $Lm1 = Lm - 2S_0 + FAI(t)$ and the value of the function of the area of intersection at the site from t_1 to t_2, respectively will be equal.

$$FAI(t) = Lm1 - Lm + 2S_0.$$

This model for obtaining FAI of no raster images is merely exemplary in the possibility of its implementation. This paper considers mainly the variant of raster image processing. The ability to handle no bitmaps or images with an irregular arrangement of the elements is one of the key advantages of the application of methods based on parallel shift technology.

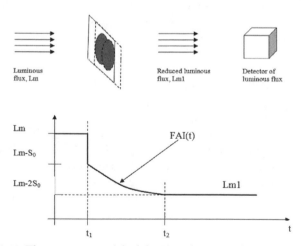

Fig. 3.12: The process model of the FAI obtaining for not bitmaps

4

Methods of Dynamic Images Processing Based on PST

Modern machine vision systems involve the use of processing of video data in real time. In particular, as example of such systems, can be autonomous robots, assembly lines in manufacturing and surveillance systems (Belan and Yuzhakov, 2013). The development of machine vision systems can also be the basis for the human eye prosthesis.

The main tasks that are solved with the help of machine vision systems are the following: Identification of the object of observation, determination of its spatial location, physical parameters and parameters of movement in space.

Recognition involves identifying an affiliation of object to a class, which is the basis of the action analysis of intellectual system for further interaction with the object (Chen et al., 1995). Furthermore, in automated assembly lines, this task includes the process of determining existence of specified properties and the absence (or presence) which causes an assignment of the product.

Modeling of the spatial position of objects is carried out for two coordinate systems (internal and external). Location of objects in the external coordinate system allows creation of a model of the world. This model exists objectively and does not depend on the actual location of IS. The outer world model is suitable for use in many intelligent systems, regardless of their individual characteristics. Geographic maps can be a simple example of external models of the world. The construction of these models is easiest to do with the help of external positioning devices. So an important achievement in this area is the use of GPS systems. However, the establishment of external models of the world has no relation to the parallel shift technology. Therefore, this aspect of video processing is not considered in this book.

Parallel shift technology allows processing of the objects that can be perceived by the system as a flat image. These objects should be in immediate surroundings with IS. Defining location relative to the internal coordinate system is necessary to carry out by an intelligent system the necessary acts of interaction with the surrounding objects. This model is subjective and individual for each IS.

Determination of the physical parameters (linear dimension, area, volume, contour shape) is inextricably linked with the processes of identification and spatial localisation. This process allows replenishment of the environment models of real objects.

The main characteristics of object motion parameter are direction, distance and speed of displacement. Some systems use the parameter as acceleration, but it is just a derivative of the velocity. Therefore, it is not related to the basic parameters of motion. Knowledge of motion parameters gives the intelligent system the opportunity to construct the algorithm of the following actions in time. This becomes one of the main tasks of intelligence-prediction of the future for a timely response to the impact of external factors.

Hardware implementation of video surveillance process is based on the use of system's 'sensor-analysis unit'. The system uses various types of sensors. Basically it is usual (or special) video camera, but can also use ultrasonic echolocation devices and other devices to signal the spatial location of the object. The number and type of sensors are selected on the basis of the need to achieve the goal which is implemented by this intelligent system and its geometric parameters.

The difference in ordinary and specialised cameras is that a first-class camera fixes the image for further transmission to the analysis unit. The second class of cameras has built-in processor system, which transforms images in a certain way that allows it to accelerate further analysis.

The video surveillance systems can be part of both fixed and mobile intelligent systems. The advantage of stationary systems is the ability to use a wide range of different types of sensors and the possibility of their location in free space. Biological beings, autonomous robots and video surveillance systems in transport belong to moving objects. Use of sensors in these cases is limited by the size of the organism or the technical means. To determine the spatial location and the physical parameters of external objects, it is advisable to use a binocular video system, laser sensors and devices like ultrasonic echolocation.

An important factor in image processing is the speed of information exchange within the video surveillance system. Obviously, at the stages of perception and preliminary preparation of data for further analysis, the system built on the basis of PST operates more slowly than in the case of direct data transmission to the analysis unit. The process of forming a set of functions of the area of intersection leads to certain time expenses.

However, adoption of the final decision on the classification of the object in this case is easier to implement. The use of parallel shift technology allows regularisation of information. Any reduction of entropy leads to an increase in the productivity of the intellectual systems.

4.1 Determination of Spatial Orientation of Planes

Common image recognition algorithm based on the use of parallel shift technology has been previously described in the book. Besides pattern recognition, this algorithm with minor modifications can be used to determine the spatial orientation of the object. In this case, the object is considered as a flat image. In the three-dimensional object, its surface is composed of many areas, each belonging to a separate plane and as a flat image. If there is an opportunity to select these areas of the common image, it is possible to determine their spatial orientation (Belhumeur et al., 1997).

Let us assume that we have an image, the parameters of which are stored in memory of recognition device as A pattern. At an arbitrary location of the image in the space of this figure will be displayed as a B projection of the A pattern to receptor field plane (Fig. 4.1).

When determining the spatial orientation of the plane with the figure, the property of the projection area is used.

$$S_B = S_A \cos \gamma \tag{4.1}$$

where γ is the angle between the planes, which contain the A and B figures. These planes for sake of convenience will be called analogically A plane and B plane.

B plane is located perpendicularly from the intellectual system to the image. The peak of plane γ angle is a straight line that forms in the plane of the receptive field (B plane) and the φ angle with the horizontal. When

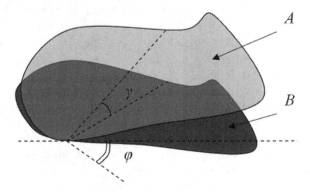

Fig. 4.1: The projection B for etalon A to arbitrary plane

shifting copies of the figures A and B along this line, the corresponding values of FAI for this area are linked by the property of the projection area.

It is not possible to immediately determine the φ direction. It is, therefore, necessary to carry out a search for all shifts of the B figure copies in the direction from 0 to π. Conducting a search for directions from π to 0 is not necessary due to the presence of central symmetry in the sets of the function of the area of intersection and integral coefficients. This procedure is similar to the rapid recognition process (a common recognition algorithm described earlier). It compares the integral factors of the real B image for directions $x = \varphi$ and $y = \varphi + \dfrac{\pi}{2}$. At this stage, it is used for integral coefficients, the constant property when changing the basic parameters of the functions of the area of intersection, i.e. when scaling.

Thus, to determine the φ direction, it is necessary to perform the following. For each of the possible directions of displacement in the range between 0 and π we get the function of the area of intersection and their integral coefficients of the real image. The Y class etalons are determined, which satisfy the following formulas:

$$k_{x_A} = k_{x_B} \tag{4.2}$$

$$k_{y_A} = k_{y_B} \tag{4.3}$$

For the class Y figures, it is necessary to determine which of the directions $x = \varphi$ or $y = \varphi + \dfrac{\pi}{2}$ is the direction of the line—the top of the plane γ angle or the direction of the projection.

To find the right projection direction, it is essential to compare the function of the area of intersection of real images and the function of the area of intersection of Y class etalons for directions $x = \varphi$ and $y = \varphi + \dfrac{\pi}{2}$. This is similar to the detailed stage of the recognition algorithm. The comparison is made by taking into account the properties of the projection area (4.1). Since the dimensions of the pattern figures and the real image may differ, the scaling of corresponding FAI is applied. Before comparing, scaling with preservation of function proportions is performed.

On this stage, the following four possible scenarios are possible:
Option 1.

$$\frac{FAI(x)_B}{FAI(x)_A} = \text{const} = \cos \gamma \tag{4.4}$$

and

$$\frac{FAI(y)_B}{FAI(y)_A} = \text{const} > 1. \tag{4.5}$$

If the formulas (4.4) and (4.5) are true, then this pattern is selected correctly. Projection takes place relative to the direction of perspective $x = \varphi$; else we move on to the next etalon of Y class for which the true formulas are (4.2) and (4.3).

Option 2.

$$\frac{FAI(x)_B}{FAI(x)_A} = \text{const} > 1 \tag{4.6}$$

and

$$\frac{FAI(y)_B}{FAI(y)_A} = \text{const} = \cos \gamma \tag{4.7}$$

If the formulas (4.6) and (4.7) are true, then this model is selected correctly. Projection takes place relative to the direction of perspective $y = \varphi + \dfrac{\pi}{2}$; otherwise, we move on to the next etalon Y class for which the true formulas are (4.2) and (4.3).

Option 3.

$$\frac{FAI(x)_B}{FAI(x)_A} = \text{const} = 1 \tag{4.8}$$

and

$$\frac{FAI(y)_B}{FAI(y)_A} = \text{const} = 1 \tag{4.9}$$

If the formulas (4.8) and (4.9) are true, then this model is selected correctly. This option is fully consistent with the detailed step of the recognition algorithm. The real image B form corresponds to the standard A form; otherwise, we move on to the next etalon of Y class, for which the true formulas are (4.2) and (4.3).

Option 4.

$$\frac{FAI(x)_B}{FAI(x)_A} \neq \text{const} \tag{4.10}$$

and

$$\frac{FAI(y)_B}{FAI(y)_A} \neq \text{const} \tag{4.11}$$

In this case, the pattern is not correctly selected; we go to the next pattern of Y class, for which the true formulas are (4.2) and (4.3).

At all stages, comparison is being made with regard to the valid coefficients of comparison dk and dS.

If no pattern is found which can have projection as real *B* figure, then it is advisable to perform a system learning process for inclusion in the array of patterns.

For simple shapes, it should be considered that each triangle is a projection of an equilateral triangle, each ellipse is a projection of the circle, and each parallelogram is a projection of a square. These properties can be used in the classification of ellipses and parallelograms.

With this method was calculated angles φ and γ and we can accurately determine the spatial orientation of the 2D shape relative to the plane of the receptor field, if its parameters are available among the array of pattern. Determination of the spatial orientation of objects is necessary for the operation of automatic systems 'eye-hand' of autonomous robots and of assembly lines. No effective methods for selecting certain parts of the surface of 3D-objects have been developed. In this case some of these component parts that are irregularly allocated on the assembly line may be labelled, thus allowing capturing of components by devices of assembly line using the method for determining the spatial orientation of the planes on the basis of parallel shift technology.

4.2 Use of Different Types of Shifts of Image Copies in Video Processing Devices Based on PST

Before proceeding to the description of the methods for determining the parameters of the movement of objects on the basis of the use of the parallel shift technology, one important point should be emphasised. It was previously mentioned that it is possible for formation of the function of the area of intersection using both cyclic and non-cyclic shifts of image copies. For example, in a cyclic shift to right, the following actions take place. If a copy of the figure falls outside the right boundary of the receptor field, it appears to be at the same level as the left portion of the receptor field and continues its movement to the right (Fig. 4.2).

If not using a cyclic shift of the figure copy to the right, it goes beyond the right edge of the receptive field. Shift type is determined by the copy of the image matrix, which is built on the selected shift registers. In the section about the formation of the function of the area of intersection at the stage of pre-treatment, it was named as the *B* matrix. If it was built on the cyclic shift registers, the cyclic shift process is obtained. When using no cyclic shift registers, no cyclic shift occurs.

Different types of copy image shift may be applied in various devices. Methods of determining the object motion parameters through the use of a parallel shift technology will also differ. Organisation of the cyclic shift should be applied mainly in prosthesis of visual organs in biological beings. The biological information systems operate with uninterrupted stream of information, which is perceived by their receptors.

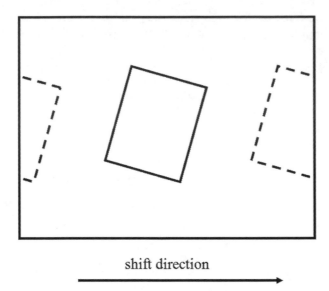

shift direction

Fig. 4.2: The process of cyclic shift of the figure

The disadvantages of using the cyclic shift of copy image are the following:

1. The inability to reach a situation of zero intersection, if the double maximum sizes of the figure in the direction of displacement occur at more than half of the width of the receptive field in the same direction.
2. *The time dependence:* To carry out certain image processing tasks it is necessary to analyse different parts of the continuous cyclic functions of the area of intersection. The process of formation of the FAI, the length of which is sufficient for analysis, takes a certain period of time, i.e., you need to spend a few cycles of the shift copy of the image for the solution of problems. When moving an object, its image may shift significantly in the receptor field area relative to its initial position. It makes much more difficult or even impossible to determine its intersection with the moving copy, but the decision may be an increase in shear rate of the image copy.
3. *The hindered process of the initial formation of the copy of a moving image:* The main disadvantage of methods based on non-cyclic shift images leads to a need for partitioning a video stream of frames at different times. It is very difficult to use this approach to the formation of the function of the area of intersection when use in non-digital IS. However, using the non-cyclic shift of the figure copies in the organisation of machine vision systems simplifies the construction.

Ways of solving these problems will be shown further in the presentation of appropriate methods of video processing.

4.3 Methods for Determining Movement Parameters Using Cyclic Shifts

The use of parallel shift technology allows determination of the motion parameters of the object. The main parameters of the movement will assume the direction and distance of the object displacement in the receptor field. Let's describe methods for determining the motion parameters of the object by using cyclic shifts of the image copies in the formation of the function of the area of intersection.

If the original image is still, the function of the area of intersection is periodic and symmetrical with respect to each of the points of complete coincidence of the input image, subject to real state, and its copy, which is being shifted (Fig. 4.3).

Here the parameters of the function of the area of intersection equals the following values: FAI period (T)—width of the receptive field in shear by horizontal and height in shear vertically; the width of each FAI splash (function portion that corresponds to a non-zero crossing)—$2X_{max}$ or $2Y_{max}$ respectively with shift direction; and the maximum FAI amplitude—the area of the input image (S_0). Orthogonal directions of displacement are selected for ease of hardware implementation. The period (T) is calculated in units of distance measurement. This is useful for raster image processing. For non-raster image, it is convenient to express this option by velocity of the copies shear (V) and time (t), for which there is a defined process.

For example, $$T = Vt_T \quad \text{or} \quad X_{max} = Vt_{X_{max}}$$

where t_T is execution time for the shift of the image copy on the width of the receptor field; and t_{Xmax} is time of displacement on the value of the maximum shift.

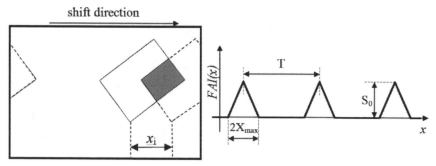

Fig. 4.3: Process image of the cyclic shift and a schematic representation of the cyclic FAI (x) graph

If the image, for which the motion parameters are determined, moves in the receptor field in the orthogonal direction, then the cyclic shift of copies in the same direction is the period (T) of function which will vary depending on the direction of motion of the figures. While moving in the direction of displacement of the copy, the period between the peaks function (call it T_1) due to the Doppler effect will be greater than T, and in the reverse direction it would be less.

Suppose the image does not move in the orthogonal direction and the copy of the shift takes place so fast that during the calculation of the intersection area, the figure has no time to be shifted in the direction perpendicular to the shift. Then each splash of cyclic FAI will be symmetric, but its amplitude is less than the fixed input image. In this case, the period of the function (T_1) is defined as the distance between the centres of two adjacent bursts with maximum amplitude.

Let us assume that a copy of the figures moves to the right in determining FAI (x) and in determining the FAI (y), it moves up and the centre of mass of the figure is at the origin of the coordinates. Assume that there is no movement in the orthogonal direction, then the tangent of the angle of the direction of motion pictures (α) will be equal to:

$$\text{tg}\alpha = \frac{\Delta T_y}{\Delta T_x} = \frac{T_{y1} - T_y}{T_{x1} - T_x}, \tag{4.12}$$

where T_{y1} is FAI (y) period for the moving image, T_y – FAI (y) period for a still image, and T_{x1} and T_x are corresponding FAI (x) periods.

When the orthogonal movement of the object,

$$\Delta T_y = T_{y1} - T_y = 0, \tag{4.13}$$

if the object moves horizontally,

$$\Delta T_x = T_{x1} - T_x = 0, \tag{4.14}$$

If the object moves vertically, while moving in the vertical movement the direction α angle is equal to $\frac{\pi}{2}$

The distance of the object displacement for one period of offset copy will be as following:

$$\Delta T = \sqrt{\Delta T_x^2 + \Delta T_y^2}. \tag{4.15}$$

It should be noted that determining time of the parameters ΔT_x and ΔT_y must be synchronised. For this purpose, in hardware implementation, the width and height parameters of the receptor field (T_x and T_y) should be same.

Signs of given parameters ΔT_x and ΔT_y and tgα uniquely determine the direction of motion of the image from its original position (Fig. 4.4).

The determination method of the basic parameters of motion is limited only by the speed of the technical capabilities of the cyclic shift of copy.

Limitations based on the properties of the orthogonal movement can be used in video surveillance systems on the road, waterway and railway transport. This facilitates manoeuvre of these types of transport in a horizontal plane. Such limitations can greatly simplify the construction of automated video surveillance systems for the determination of motion parameters of foreign objects.

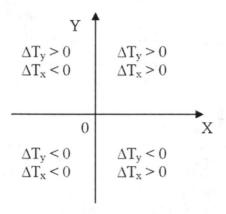

Fig. 4.4: The direction of motion of the object depending on the signs ΔT_x and ΔT_y

4.4 Methods for Determination of Motion Parameters Using Non-cyclic Shifts

The methods for determining the parameters of the object motion by using non-cyclic shifts of the image copy in the process of FAI formation are based on the determination of the centre of mass of the object. The continuous video stream is divided into frames corresponding to certain points in time. For each point in time, the position of the object in the receptor field is determined. Furthermore, the coordinates of the figure at the current moment are compared with similar coordinates at the time corresponding to the previous frame.

In digital processing of information, the coordinates of the masses centre of the image (x_{cm} and y_{cm}) are defined as the average value of the coordinates of all its n elements.

$$x_{cm} = \frac{\sum_{i=1}^{n} x_i}{n} \tag{4.16}$$

$$y_{cm} = \frac{\sum\limits_{i=1}^{n} y_i}{n} \qquad (4.17)$$

Obviously, the number of operations when determining the coordinate of the mass centre of image increases with the number of its elements. In addition, it is not clear how to determine the coordinates in the study of no bitmap image or an object with an irregular location of the elements.

Using parallel shift technology, the image coordinates of the centre of mass can be determined as follows. When shifting the copies of the figure in the orthogonal direction it is necessary to fix the time when the area of a part of the figure copy, which is located in the receptor field, will be equal to half the area of the initial image $\left(\dfrac{S_0}{2}\right)$. The shift value at this time will be one of the coordinates of the centre of the mass (Fig. 4.5). The value of the coordinates of the centre of the mass in the other orthogonal direction is defined similarly.

It is noted that in this case, a coordinate means the distance from the centre of the figure mass to the receptor field boundary. Binding to the point of origin is absent. When fixing the origin at a certain spot, you need to carry out appropriate corrections of coordinates in the centre of mass figure. For example, when fixing the start coordinate in the upper left corner of the receptive field, the corrected coordinates of the mass centre of the image (x'_{cm} and y'_{cm}) will be as follows.

$$x'_{cm} = T_x - x_{cm}, \qquad (4.18)$$

$$y'_{cm} = T_y - y_{cm}. \qquad (4.19)$$

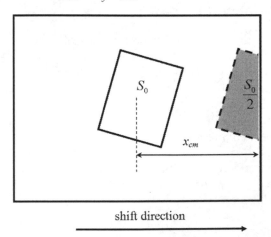

shift direction

Fig. 4.5: Graphical display of the process of determining the x_{cm} coordinates of the mass centre of the image

The distance and direction of the object displacement in the receptor field are determined by analysis of the adjusted coordinates of mass centre of the image for each pair of adjacent frames.

4.5 Methods for Determining Distance of a Moving Object

In the previous sections the methods that allow definition of the parameters motion of the object in a plane that is perpendicular to the direction of view are described. These methods do not allow determination of the distance to the studied image. The following is a method for determining the distance to an object by analysing the function of the area of intersection.

For spatial localisation of the object, it is necessary to determine the distance to it. Definitions of distances to an object can be done in many ways, for example, using an ultrasonic echolocation, laser and the binocular vision system building. In order not to apply the other character technology, PST is being used for determining this parameter. Determination of the distance to the 2D object can be performed by using the similarity properties of Fig. 4.6.

According to these properties

$$\frac{l+\Delta l}{l} = \frac{Y_{max\,0}}{Y_{max\,1}} = \sqrt{\frac{S_0}{S_1}} \,, \tag{4.20}$$

wherefrom

$$l = \frac{\Delta l \sqrt{S_1}}{\sqrt{S_0} - \sqrt{S_1}} \,. \tag{4.21}$$

where l – the initial distance to an object; Y_{max0} and S_0 – FAI(y) parameters at time t_0; l_1 – distance to the object; Y_{max1} and S_1 – FAI(y) parameters at time t_1.

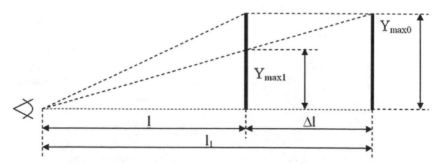

Fig. 4.6: Graphic model of the process of determining the distance to the object

To determine the initial object distance l can be used by changing the distance to the object by moving the optical system itself or by changing the focal length of the perceiving camcorder (similar to the process of accommodation of the eye). Thus, value $\Delta l = l_1 - l$ is specified. For obtaining values $\sqrt{S_0}$ and $\sqrt{S_1}$ it is necessary to determine the object area in the moments before (t_0) and after (t_1) by changing the focal length or the distance of displacement of the optical system.

In the proposed method, t_0 and t_1 moments correspond to the establishment of two neighbouring frames when no cyclic shift of a copy image occurs.

During the cyclic shift copies of the image owing to changes in the distance to the object, the shape of continuous functions of the area of intersection will change. When removing an object from the observer, a form of another FAI surge will be smaller in size ($S_1 < S_0$), its width will change a little ($X_{\max 0} \approx X_{\max 1}$) and at the top part, the horizontal component will appear. When an object approaches the observer a form of regular FAI surge changes as follows. The maximum FAI value will not change ($S_1 = S_0$), but its width will change significantly ($X_{\max 0} < X_{\max 1}$) and at the top the horizontal component will appear. Examples of changing the shape of the cyclic function of the area of intersection are shown in Fig. 4.7.

The distance of the object in this case should be analysed by two parameters areas (S_1 and S_0) and the maximum shift values ($X_{\max 0}$ and $X_{\max 1}$), to use those that vary. When removing the figure in the calculation, we use the value of the function of the area of intersection.

$$l = \frac{\Delta l \sqrt{S_1}}{\sqrt{S_0} - \sqrt{S_1}}$$

When approaching a figure in the calculation, the maximum shift value is used.

$$l = \frac{\Delta l \cdot X_{\max 1}}{X_{\max 0} - X_{\max 1}}$$

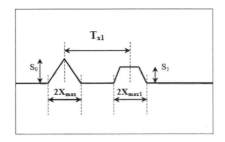 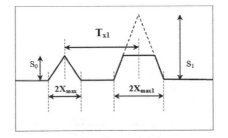

a) deleting of object   b) object approaching

Fig. 4.7: Examples of the shape changing the cyclic FAI

Determination of the distance to the object by given methods is limited to the possibility of organisation of the processes that determine the area of the image and rapid parallel shifts of the copies. When using a cyclic shift of a copy of the image, a shift speed copy is very important. The larger it is, the more accurately we can determine the object location.

4.6 Comparison of Methods for Determination of Motion Parameters by Using Different Types of Shifts of the Image Copy

To compare methods of determining motion parameters by using different types of shifts of copy of image, the software model was created. Three types of trajectories of the moving object were tracked. The first type—the position of the point of intersection of the diagonals of the bounding box is determined; the second type—a location of the mass centre of the figure is determined; and the third type—the calculation of trajectory by analysing the cyclic FAI is used.

The video file of motion black object was used for the tests on a white background in AVI format with resolution 640×480. The sequence of individual frames of the text of the video file is shown in Fig. 4.8.

The result of the images of trajectories of the object motion of three types is shown in Fig. 4.9.

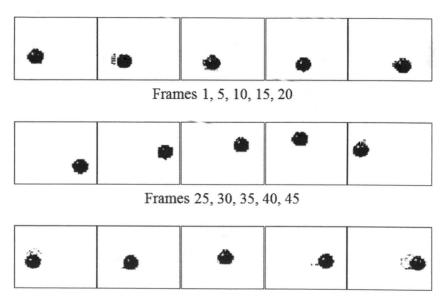

Frames 1, 5, 10, 15, 20

Frames 25, 30, 35, 40, 45

Frames 50, 55, 60, 65, 70

Fig. 4.8: The sequence of individual frames of the test video

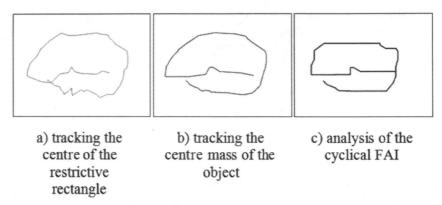

| a) tracking the centre of the restrictive rectangle | b) tracking the centre mass of the object | c) analysis of the cyclical FAI |

Fig. 4.9: Image of different types of trajectories of motion of the object

When determining the position, the point of intersection of the diagonals of the bounding box has a distortion due to changes in the shape of the test object. If the image preprocessing is performed for each shift, the trajectory would be smoother. The results of this method are shown only for control and are unrelated to the PST.

Simulating the definition of the path of movement by analysing the cyclic FAI is most difficult. To determine periods (T_{x1} and T_{y1}) we had to look for peaks in certain areas of continuous functions of the area of intersection. Besides the synchronisation process to determine the function of the area of intersection was difficult because of the different sizes of width and height of the receptor field. Resolving capacity of 640×480 was selected for the possibility of creating a test video file by standard means.

The method of tracking the centre of masses of the figure in software implementation was most effective. This method is suitable for implementation of machine vision systems.

Despite existing disadvantages in general, there is an obvious correlation between the data obtained by different methods. The main advantage of the methods for determining the parameters of movement of objects on the basis of parallel shift technology is the ability to handle video data with no digital image processing.

5

Image Processing System Based on PST

On the basis of the proposed methods, which use parallel shift technology to implement a variety of image processing tasks, you can build a computer vision system. Devices constructed with its application can be used as an independent complex or as part of data processing of the hybrid system with digital image processing devices.

In general, the intelligent systems operate by a scheme of 'receptor—analysis unit—effector'. In the process of information processing in the form of 'receptors' and 'effectors' simple sensors and manipulators are used and they do not perform any analytical operations. The analysis unit in this case is the processor. To improve the performance of IS specialised devices of perception are developed and the impact on the object seen as by itself begins to perform some analytic functions. Circuits of interaction of elements of the intelligent system in the implementation of tasks of perception, analysis and cooperation in the use of simple and specialised external devices are shown in Fig. 5.1.

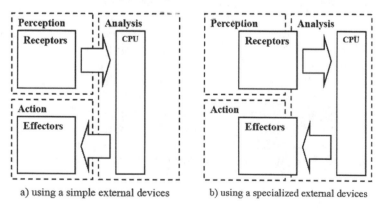

a) using a simple external devices   b) using a specialized external devices

Fig. 5.1: Schemes for interaction of elements of intelligent system

5.1 System of Perception of Video Information

Parallel shift technology allows transformation of the information to improve the performance of the units it is processing. The result of the use of parallel shift technology is generalised information that will simplify the processing of the data. In this book is described the possibility of using this technology for construction of specialised video-processing devices.

In the above methods of information processing, manipulation with binary images is considered. Furthermore, the possibility of creation of no digital devices at this time is not sufficiently studied. Only the study of analog computing devices is undertaken. Due to these facts the video system of perception should include the steps of digitising and binarisation. Also it is necessary to have tools for noise control. The main blocks of the image-processing devices based on parallel shift technology are the blocks of the function of the area of intersection formation and of analysis of these functions. The structural circuit of one of the image processing device options is presented in Fig. 5.2.

Methods of dealing with noise and the formation of functions of the area of intersection are considered. Special attention is paid to the process of image bundle at this stage. There are several methods for image separation in certain areas. Morphological separation consists of selection of connected regions. However, to organise such a division by hardware methods is difficult. Therefore, at this stage it is possible to determine the quantitative characteristics of dividing the object into sections depending on the brightness of its elements. Splitting of images by morphological characteristics can be done during the analysis phase.

For each biological species there is a certain range of parameters of the environment in which the possible perception of this or that sensory organ occurs. For example, a person perceives the frequency range of sound waves from 4 Hz to 16,000 Hz, and visually perceives the frequency waveband from 384 to 790 THz. The sensitivity of the sense organs to quantitative characteristics of the change is also limited. Waves with close rates of frequency values are perceived as equal. For hardware implementation of the separation process on the image portions, precise quantification of such areas is required. Close brightness values of the image elements fall into one group. This process simulates the perception of the environment by biological intelligent systems.

The image may be divided into ranges in different ways. Figure 5.3 shows three possible variants of division of the input object in dependence on the ratio of the specific threshold values of brightness. For further processing all the options of bundle simultaneously may be used.

Assume that the input image is divided into (m) number of sections according to the brightness gradations. Then each of them will be a separate image. For each such layer it is necessary to define a set of the

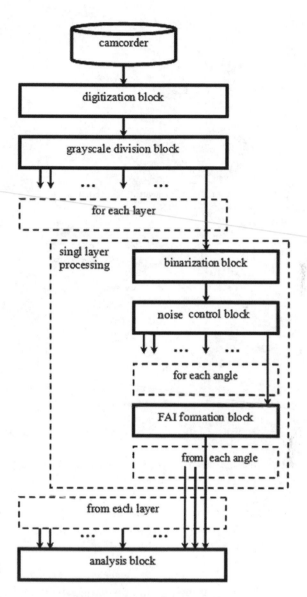

Fig. 5.2: Structural diagram of one embodiment of image-processing device

function of the area of intersection. The process for each layer necessarily includes binarisation and noise control means. The process of removing noise components was previously described. Further, each of the m images is sent to the n device of construction of the functions of the area of intersection. This number of devices is determined by the number n of

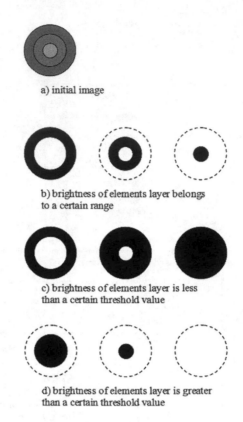

a) initial image

b) brightness of elements layer belongs
to a certain range

c) brightness of elements layer is less
than a certain threshold value

d) brightness of elements layer is greater
than a certain threshold value

Fig. 5.3: Examples of dividing an input image into segments

directions, which are determined by FAI. During software modelling the image $n = 36$ was used. Thus, the range from 0° to 180° with a step 5° was covered.

As a result of device work on the unit of analysis is supplied a set of $m \times n$ functions of the area of intersection for further processing. In the analysis unit image processing methods that are described earlier are implemented.

The presented version of an image processing device is more suitable for biological intelligent systems. In this case, the FAI generating unit may include a set of registers of cyclic shift of the copy image. Information flows from the camcorder to the digitisation unit and from the formation of FAI block on analysis unit are continuous. They are not divided into individual frames.

The scheme in image processing device for use in computer systems somewhat changes (Fig. 5.4). After the digitisation process is added the framing unit. We assume that the frame is a slice of video stream

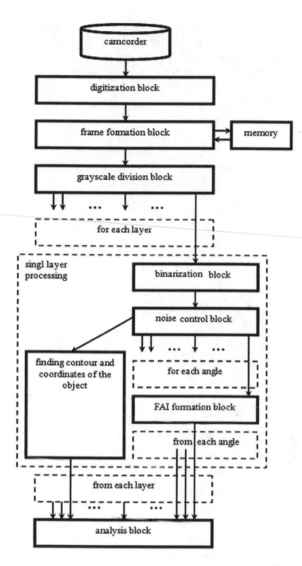

Fig. 5.4: Structural diagram of an image-processing device with preliminary selection of the frames from video stream

at a certain time. The frames may be stored in memory for subsequent processing by the same device or for transmission of manipulations in another data processing system; for example, for digital image processing.

In the FAI case, the set of cyclic shift registers is not necessarily used. Simultaneously with the formation of functions of the area of intersection edge of the object can be selected to determine its coordinates. Then the analysis unit will need to transfer not only the $m \times n$ values of the functions of the area of intersection, but also the images of m areas edge of the object and $2m$ values of their coordinates.

For the processing of objects with dimensions of more than two in such devices, the camera must be replaced by a corresponding detector. The forming units of intersection object functions and its copy must also be modified.

In the analysis of continuous functions, it is necessary to know what areas of FAI graphics is responsible for performing certain operations. In Fig. 5.5 certain functions of the area of intersection with the specified areas are and the analysis of which allows performing certain image-processing tasks (Dai and Nakano, 1998).

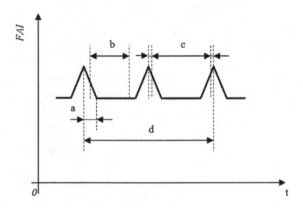

Fig. 5.5: Distribution of parts of the FAI graph to perform certain operations

The a graph segment is used to implement the pattern recognition process for determining the spatial orientation of objects—segment b—in determination of the coordinates of the initial location, area c—in determining the outline of the image and in the fight against noise, area d—in determination of the parameters of the object. The FAI graph on Fig. 5.5 is schematically shown and only in one displacement direction of the image copy. The right boundary of b segment is chosen arbitrarily. It depends on the actual distance from the centre of mass of the object to the edge of the field receptor. The FAI area specialisation for other areas is similar. Here the function of the area of intersection depends on the time (t). In order to determine the function values dependent on the distance, it is sufficient to multiply the corresponding time periods on a shear rate of object copy.

The need to take into account the Euclidean coordinate system, some of the image-processing tasks are enough to only perform orthogonal directions shear copy. These tasks include determination of initial coordinates of an object and its movement parameters.

To perform each of these tasks, a separate device is constructed by using non-cyclical FAI. These devices are included in the image-processing system in parallel with the process of formation of the function of the area of intersection. Determination of movement parameters is performed by calculating the centre of mass of the object.

Based on the proposed image-processing methods, use of parallel shift technology can construct an image-processing system. The block diagram of this system is shown in Fig. 5.6.

The proposed image-processing mechanisms in some cases can be more productive than in the use of digital processing. The presence of such devices in the machine vision systems allows organisation of rapid

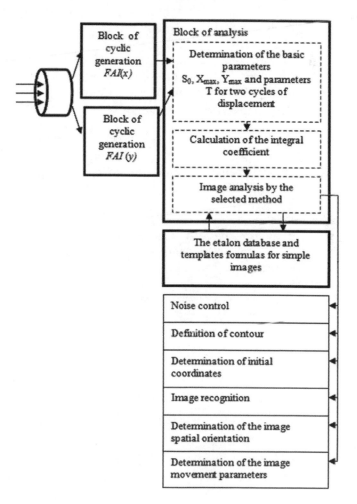

Fig. 5.6: Scheme of machine vision system

reaction processes to external factors. However, the use of parallel shift technology for information processing gives advantage only when a hardware implementation of such a system is used.

5.2 Saving Template Functions of the Area of Intersection

The analysis blocks for all variants of imaging devices should include memory units. For each image it is necessary to save a certain pattern surface which consists of the given FAI figure for $\overline{0,\pi}$ directions of shift. It will allow the implementation of tasks of recognition of images and determination of their spatial orientation. Save the pattern functions for all directions $\overline{0,2\pi}$ is not required, as the reference surface has a central symmetry.

Only a set of FAI reflects in itself information about the form of a specific object. Whatever additional factors are not used in the image-processing system to speed up the action of processing algorithms, at the final stage it is always the process of comparing the actual and pattern FAI.

Arrays of template surfaces can be stored in digital or in analog forms. Saving of standard surfaces in digital format entails numerous hardware costs and the availability of additional processing means in the comparison of functions. Before comparing, it is necessary to scale the real and etalon FAI.

There are several options to reduce the amount of necessary information for storage of pattern surfaces; for example, to store in the database of reduced copies of the template images. In this case, the value of standard FAI is generated from a template in the process of recognition, but in increased recognition time, because for each stored template, it is needed to generate the pattern surface at each recognition. Stored etalons must be small enough to reduce the hardware cost of storing this information. Scaling of standard functions towards increase will lead to appearance of false information. Comparison of functions will be quite rough, leading to errors in recognition.

Saving of pattern surfaces in analog form can, with high precision, display the functions of the area of intersection. A variant of this saving may be the use of holography. The holographic data storage methods are at research stage.

Another method of saving the templates is to replace of the pattern FAI by a certain auxiliary function that depends on the input function of the area of intersection. The rapid transformation of the function in a Fourier series was used in software simulation of image processing on the basis of parallel shift technology. The displacement range from zero to the maximum value of the shift in the chosen direction was used. This option is simple in the process of implementation and significantly reduces the hardware costs and does not require additional function-processing methods.

The amount of information needed to store template surface as elements of auxiliary function depends on the number of templates in the database.

Based on PST, image processing encourages the development of new methods of information storage. The set of actions for the preparation of comparison of templates and of real FAI at various variants of saving of standard surfaces is shown in Fig. 5.7.

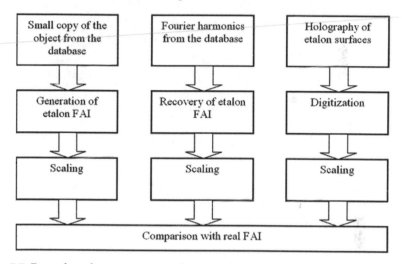

Fig. 5.7: Procedure for preparation of comparison etalon and real FAI in different variants of saving etalon surfaces

A possible solution to the problem of saving standard FAI is the use of their transformation into a Fourier series. As is known, such a transformation brings full information about the function. However, the use in processes of storage and restoration of data of Fourier series, which describes the function of the area of intersection, no given advantage is seen in reducing the amount of stored information. For this, storage may be used only in part of the harmonics of series that is similar to the low-pass filtering process.

In addition, the value of the harmonics of Fourier series depends on the area of the pattern figure. For unification the size of the harmonic values needs to be normalised. To do this, the value of each element of a number should be divided by the value of the input image area. In this case, the values of all harmonics will be close to unity, which simplifies the storage of information in databases.

In constructing the Fourier series, which reflect the standard FAI, the following formula of normalized harmonics (Mills, 2006) was used:

$$FFTN_0 = \frac{1}{S_0} \frac{1}{\sqrt{X_{max}}} \sum_{j=0}^{X_{max}-1} FAI(j) \tag{5.1}$$

$$FFTN_i = \frac{1}{S_0} \frac{\sqrt{2}}{\sqrt{X_{max}}} \sum_{j=0}^{X_{max}-1} FAI(j) \cdot \cos\left(\frac{\pi \cdot i \cdot (2 \cdot j + 1)}{2 \cdot X_{max}}\right) \tag{5.2}$$

The value of the initial image area (S_0) and the maximum shift (X_{max}) are the basic parameters of the function of the area of intersection. Reverse recovery FAI is implemented according to the formula (5.3):

$$FAI(j) = S_0 \cdot \left(\frac{FFTN_0}{\sqrt{X_{max}}} + \frac{\sqrt{2}}{\sqrt{X_{max}}} \sum_{i=1}^{X_{max}-1} FFTN_i \cdot \cos\left(\frac{\pi \cdot i \cdot (2 \cdot j + 1)}{2 \cdot X_{max}}\right)\right) \tag{5.3}$$

The data set (S_0, X_{max}, $FFTN_0 \div FFTN_i$) is taken from a pattern database.

Example of saving the pattern surface of the image of the Chinese character 'dragon' in the form of the initial twenty harmonics of the Fourier series is shown in Fig. 5.8.

In the simulation of the process of saving standard surfaces (Belan and Yuzhakov, 2014) normalised harmonics of the Fourier series ($FFTN_0 \div FFTN_{19}$) were used for 36 directions (ANGLE) in steps of 5 degrees. The directions ranging from 0 to π were gripped by the need to save the pattern surface. Value of the initial area of the real image (S_0) is also stored and the maximum displacement in the selected direction (X_MAX) was obtained in the learning process of the recognition system. The value of these two parameters corresponds to the data of the real image of the learning sample. They are used in the process of restoration of functions of the area of intersection for some additional coefficients and in scaling. Because in modelling digital images are used, the maximum displacement is calculated in pixels and the initial area is equal to the number of pixels belonging to the image. WHAT field contains the name of the object, which is given by man; field INTEGR is the value of the integral coefficient for the corresponding direction. It is used for quickening image processing in the various processes.

Creation of a software model of the process was intended to demonstrate image processing on the basis of the parallel shift technology. Therefore, the selected structure of the pattern database is not perfect. In the case of practical application, it can be normalised. Also, the automatic naming of patterns may be implemented (in the WHAT field). It may be both arbitrary or generated by analysing an existing database. But that is another topic relating to the machine learning processes, which are described in other scientific sources. Therefore, we will not describe these issues in detail.

ID	WHAT	ANGLE	INTEGR	X_MAX	S_0	FFTN0	FFTN1	FFTN2	FFTN3	FFTN4	FFTN5	FFTN6
1	dragon	0	0.2484	223	25077	3.7434	2.7350	0.3868	0.7211	0.7966	0.4466	0.3314
2	dragon	5	0.2403	211	25077	3.5249	2.5884	0.5322	0.7608	0.5743	0.4148	0.3246
3	dragon	10	0.2303	220	25077	3.4502	2.7628	0.5828	0.6247	0.6335	0.4252	0.3039
4	dragon	15	0.2296	228	25077	3.5002	2.7830	0.5509	0.6241	0.6984	0.4244	0.2970
5	dragon	20	0.2364	209	25077	3.4523	2.6366	0.5487	0.8016	0.5889	0.3759	0.3340
6	dragon	25	0.2281	217	25077	3.3940	2.4593	0.6216	0.9311	0.6195	0.3734	0.3415
7	dragon	30	0.2337	207	25077	3.4836	2.3288	0.5683	1.0049	0.5860	0.3950	0.3128
8	dragon	35	0.2234	223	25077	3.3693	2.4404	0.6555	0.8800	0.5517	0.4929	0.3849
9	dragon	40	0.2139	238	25077	3.3322	2.5023	0.7118	0.8625	0.5131	0.3756	0.4306
10	dragon	45	0.2113	244	25077	3.3328	2.5684	0.7555	0.7820	0.4950	0.3417	0.4160
11	dragon	50	0.2095	251	25077	3.3502	2.5929	0.7748	0.8144	0.4287	0.3149	0.4816
12	dragon	55	0.2130	254	25077	3.4266	2.5845	0.7840	0.8178	0.3735	0.4440	0.4659
13	dragon	60	0.2159	250	25077	3.4457	2.6092	0.8301	0.7174	0.4606	0.5349	0.3472
14	dragon	65	0.2063	258	25077	3.3451	2.7951	0.8633	0.6833	0.5937	0.4864	0.2780
15	dragon	70	0.2219	246	25077	3.5127	2.7654	0.8722	0.6025	0.4823	0.3099	0.2732
16	dragon	75	0.2237	254	25077	3.5970	2.9239	0.9678	0.7234	0.3631	0.3011	0.2774
17	dragon	80	0.2273	261	25077	3.7027	3.2094	0.8581	0.5554	0.4160	0.2653	0.2675
18	dragon	85	0.2744	240	25077	4.2835	3.0589	0.4227	0.6095	0.3417	0.2573	0.2463
19	dragon	90	0.2853	239	25077	4.4426	3.0247	0.3670	0.5370	0.3300	0.3187	0.2340
20	dragon	95	0.2569	258	25077	4.1569	3.0818	0.5091	0.5263	0.3016	0.3530	0.2948
21	dragon	100	0.2436	261	25077	3.9660	2.9892	0.6520	0.6396	0.3458	0.3456	0.3158
22	dragon	105	0.2378	256	25077	3.8365	2.8732	0.6012	0.7046	0.4469	0.3329	0.3004
23	dragon	110	0.2293	261	25077	3.7360	2.7901	0.6363	0.6821	0.4509	0.4035	0.3053
24	dragon	115	0.2162	272	25077	3.5957	2.7480	0.7896	0.6586	0.4059	0.4363	0.4013
25	dragon	120	0.2031	286	25077	3.4646	2.7711	0.8298	0.6680	0.4275	0.3759	0.4411
26	dragon	125	0.1946	299	25077	3.3933	2.7538	0.8748	0.6633	0.4961	0.3881	0.3185
27	dragon	130	0.1861	311	25077	3.3095	2.7131	0.9151	0.6854	0.4781	0.4531	0.2697
28	dragon	135	0.1993	286	25077	3.3399	2.5983	0.7839	0.6579	0.4877	0.4479	0.2232
29	dragon	140	0.1305	292	25077	3.2839	2.6795	0.7660	0.5759	0.4959	0.5244	0.3006
30	dragon	145	0.1999	274	25077	3.3398	2.5920	0.6960	0.5335	0.5514	0.5586	0.2808
31	dragon	150	0.2004	269	25077	3.3179	2.5799	0.6767	0.5031	0.6510	0.5067	0.2937
32	dragon	155	0.1851	289	25077	3.1756	2.7355	0.7241	0.4505	0.6636	0.5738	0.3902
33	dragon	160	0.1343	272	25077	3.2344	2.7315	0.6010	0.5011	0.7024	0.5885	0.3632
34	dragon	165	0.2203	240	25077	3.4456	2.5331	0.4565	0.6753	0.6771	0.5303	0.2768
35	dragon	170	0.2276	230	25077	3.4842	2.5326	0.4477	0.6563	0.6972	0.4897	0.3528
36	dragon	175	0.2347	228	25077	3.5766	2.5677	0.3909	0.6485	0.7297	0.5390	0.3550

FFTN7	FFTN8	FFTN9	FFTN10	FFTN11	FFTN12	FFTN13	FFTN14	FFTN15	FFTN16	FFTN17	FFTN18	FFTN19
0.2356	0.1947	0.3951	0.2889	0.1578	0.1176	0.1259	0.0837	0.0572	0.0582	0.0616	0.0657	0.0423
0.2099	0.3579	0.3911	0.2111	0.1517	0.1228	0.1243	0.0549	0.0691	0.0895	0.0780	0.0500	0.0382
0.2553	0.3343	0.4126	0.2415	0.1304	0.1406	0.1173	0.0916	0.0572	0.0762	0.0899	0.0568	0.0441
0.2964	0.3109	0.4101	0.2852	0.1280	0.1349	0.1168	0.0773	0.0929	0.0864	0.0785	0.0630	0.0411
0.2947	0.4376	0.2998	0.1337	0.1387	0.1148	0.0966	0.0952	0.0924	0.0659	0.0515	0.0429	0.0307
0.3375	0.4066	0.3218	0.1707	0.1409	0.1172	0.1115	0.0942	0.0892	0.0800	0.0534	0.0392	0.0320
0.4539	0.3814	0.2363	0.1609	0.1339	0.1195	0.0930	0.0958	0.0751	0.0599	0.0425	0.0369	0.0299
0.4042	0.4234	0.2717	0.2055	0.1776	0.1319	0.1028	0.0990	0.0824	0.0618	0.0588	0.0366	0.0427
0.5429	0.4317	0.2534	0.2397	0.2238	0.1672	0.1317	0.0505	0.0888	0.0716	0.0565	0.0517	0.0464
0.5636	0.4319	0.3110	0.2326	0.2210	0.2009	0.1533	0.1123	0.0889	0.0543	0.0586	0.0585	0.0453
0.5107	0.3367	0.3110	0.3054	0.2558	0.2183	0.1749	0.1220	0.1034	0.0709	0.0521	0.0585	0.0561
0.3667	0.3616	0.2275	0.2604	0.3351	0.2527	0.1951	0.1290	0.1040	0.0791	0.0717	0.0684	0.0480
0.3095	0.2655	0.2726	0.2304	0.2835	0.2843	0.2197	0.1489	0.1098	0.0868	0.0637	0.0586	0.0544
0.2379	0.2623	0.2396	0.2215	0.2158	0.2436	0.2774	0.2161	0.1559	0.1083	0.0727	0.0469	0.0552
0.2459	0.2005	0.2544	0.2004	0.2314	0.2184	0.2420	0.2155	0.1686	0.0916	0.0557	0.0525	0.0487
0.2122	0.1761	0.2351	0.2235	0.2223	0.1933	0.1771	0.2012	0.2408	0.1703	0.0882	0.0463	0.0418
0.1934	0.1616	0.2199	0.2453	0.2074	0.1754	0.1647	0.1610	0.2058	0.2354	0.1628	0.0739	0.0452
0.1717	0.2008	0.2607	0.2085	0.1652	0.1768	0.1525	0.2017	0.2409	0.1651	0.0895	0.0402	0.0331
0.2038	0.1852	0.2621	0.1616	0.1878	0.1764	0.1371	0.1927	0.2192	0.1717	0.1105	0.0728	0.0542
0.2341	0.1543	0.2514	0.2132	0.1311	0.1685	0.1843	0.1209	0.1661	0.2005	0.1799	0.1403	0.0837
0.2074	0.2035	0.2648	0.2133	0.1266	0.1745	0.2105	0.1275	0.1341	0.1732	0.1860	0.1672	0.1015
0.2561	0.2173	0.2312	0.2016	0.1555	0.2154	0.1682	0.1269	0.1449	0.1470	0.1892	0.1612	0.0916
0.3260	0.2070	0.2259	0.1805	0.1785	0.2339	0.1971	0.1217	0.1363	0.1416	0.1832	0.1765	0.1088
0.3233	0.2544	0.2486	0.1615	0.1858	0.2266	0.2338	0.1417	0.1228	0.1158	0.1705	0.1797	0.1407
0.3498	0.2989	0.3005	0.1875	0.1676	0.2071	0.2737	0.1516	0.1321	0.1258	0.1500	0.1597	0.1583
0.3540	0.3688	0.3640	0.2610	0.1500	0.1918	0.2980	0.1636	0.1474	0.1232	0.1337	0.1587	0.1747
0.2857	0.3812	0.4510	0.3408	0.1686	0.2286	0.2829	0.1829	0.1417	0.1073	0.1409	0.1678	0.1732
0.3473	0.4654	0.4066	0.2872	0.3081	0.2410	0.1521	0.1196	0.1466	0.1791	0.1881	0.1369	0.1110
0.2839	0.3593	0.4365	0.4298	0.3160	0.2316	0.1469	0.1447	0.1609	0.1894	0.1616	0.1163	0.1082
0.3087	0.3688	0.4735	0.4165	0.2382	0.1763	0.1973	0.1992	0.1701	0.1201	0.1135	0.0847	0.0797
0.3738	0.2935	0.4589	0.3589	0.2823	0.2337	0.2266	0.1542	0.1284	0.1159	0.0805	0.0736	0.0920
0.3071	0.2846	0.3419	0.2892	0.3376	0.3806	0.2365	0.1509	0.1411	0.1279	0.0854	0.0708	0.0832
0.2671	0.2806	0.2921	0.4063	0.4403	0.2559	0.1688	0.1390	0.1095	0.0850	0.0906	0.0980	0.0616
0.3090	0.3574	0.3726	0.3840	0.2677	0.1489	0.1041	0.1070	0.1137	0.0957	0.0634	0.0711	0.0878
0.3522	0.2160	0.3844	0.3957	0.2056	0.1029	0.1145	0.1318	0.0931	0.0637	0.0775	0.0875	0.0721
0.2742	0.1953	0.3386	0.3710	0.1740	0.1169	0.1465	0.1079	0.0762	0.0738	0.0804	0.0657	0.0637

Fig. 5.8: Example description of Chinese character 'dragon' in the form of normalised values of the Fourier series harmonics is seen in the etalon database

All the images used examples of simulations were binary. They could be convex, irregularly shaped or comprise several objects.

The Fourier series harmonic number depends on the amount and form of patterns. In the process of software modelling it was revealed that to create a pattern FAI of convex figures, it is better to use the first five Fourier harmonics. To create a pattern FAI of no convex figures it is enough use the first ten harmonics of the Fourier. To create a of pattern FAI of figures consisting of several objects it is enough to use the first twenty Fourier harmonics. In all these cases the deviation reconstructed FAI and the real FAI do not exceed 10 per cent of the value of the real function at each displacement step. Also, when choosing an etalon that corresponds to a specific FAI, the average deviation of the real value and the pattern value for a specific section of the FAI function may be used. Examples of graph template of functions by using different numbers of the Fourier series harmonics for the restoration of FAI are shown in Fig. 5.9.

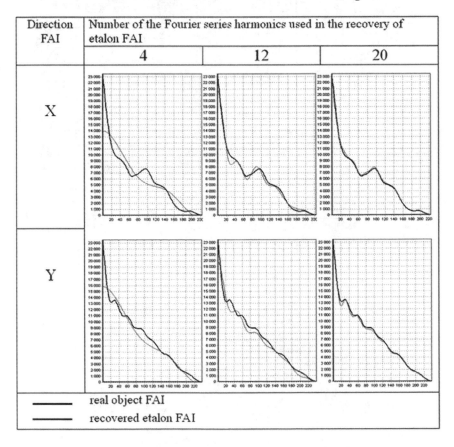

Fig. 5.9: Graphs of real object FAI and recovered etalon FAI

With the saving of pattern functions in the form of a Fourier series it is possible to simulate the process of forgetting information by man. Assume that at the initial stage (immediately after the learning process) pattern function is described by a complete set of the harmonics of Fourier series. The amount of the series of elements will be of the same length as digitised FAI. At predetermined time intervals the highest numbered harmonics from the pattern series may be removed. Some of the information about the pattern will be lost, but will be less of necessary information to store. Repeated restoration of data about the object becomes necessary when the function is built on the basis of the existing number of harmonics, and will differ from the original more than a specified value. When the selected number of harmonics of the Fourier series (NH) is for recovery of the function of the area of intersection, the formula (5.3) is converted into the following:

$$FAI(j) = S_0 \cdot \left(\frac{FFTN_0}{\sqrt{X_{max}}} + \frac{\sqrt{2}}{\sqrt{X_{max}}} \sum_{i=1}^{NH-1} FFTN_i \cdot \cos\left(\frac{\pi \cdot i \cdot (2 \cdot j + 1)}{2 \cdot X_{max}} \right) \right) \quad (5.4)$$

Saving the reference data as a function will significantly reduce the amount of data required to store. Storing information about the images as of standard surfaces will ensure implementation of tasks of recognition and determining the spatial orientation of the 2D-objects.

5.3 Application of Image-processing Methods Based on PST for Prosthetic Visual Organs

The use of parallel shift technology ensures the implementation of the main tasks of video processing and simulates the processes related with human vision. Therefore, the processing of information obtained as a result of PST can be used in medicine to create artificial bodies of view. The basis of this technology is to obtain FAI, which can be converted, for example, into audio signals. Using the human brain's ability to learn and adapt, it is possible to teach blind people to 'hear' the image and nor is it difficult to obtain the information in a suitable format for the human ear from device of cyclic generation of the function of the area of intersection. But, to determine the parameters of output signal, additional research with specialist's ophthalmologists and otolaryngologists is needed. The brightness and the amount in the scene elements may be converted to frequency and amplitude of the audio signal. In future, with the development of neurosurgery, it may be possible to give signals from receptor field directly to the visual area of the brain. In this case, the system 'sensor-analysis unit' will perform the role of the sensor device receiving FAI and the analysis unit will serve as the human brain.

In Fig. 5.10 three possible options are shown for connecting the device forming of continuous functions of the area of intersection in the human body. In the first two embodiments, the imaging device is built into the glasses and the third—in the eye prosthesis. The information is transmitted in the first embodiment in the earpiece, and in the second and third—directly on the visual area of the brain. All the three options differ in the degree of operative interference.

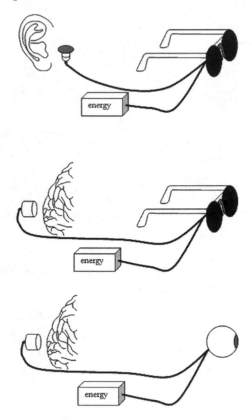

Fig. 5.10: Embodiment connections of the information-processing device to the human bodies

Research in this direction is underway since long ago, but the developers of such systems in the brain are supplied a set of pixels that make up an image. The transition information about all the elements that make up an object is quite difficult. When transmitting selected information it gets lost in perception accuracy. The image-processing systems based on parallel shift technology can use almost all the information about the visual scene. Data about areas of the image are summarized. Their number is limited, but they carry a lot of information about the object.

This machine vision system quite accurately simulates the human vision. Parallel shift technology of copy image may explain the cause of the eye pupil movement in recognition of images and edge detection of the figures. The method of calculating the basic parameters of the object allows the processes that have occurred in experiments to determine the properties of the viewed organs of living things to respond to movement. The ability to determine the angles of plane location, on which there is a figure, simplifies the operation of the system 'eye-hand'. Storage of patterns as of the pattern surfaces can explain the process of forgetting by man of the visual images. For example, if to scale the digitised pattern surface of the images that have not been used to reduce memory usage, then, when the need arises, scale them towards increasing. This way the data loss will be apparent. There is the necessity of updating this pattern. In addition, the image studied carefully, by those who have long not seen, needs to be updated in memory.

FAI graphs difference can explain how a person defines a set of letters belonging to one word (Fig. 5.11).

If an object length is equal to the maximum shift, the letters belong to the same word; else each letter is perceived separately. Then the word is formed for a set of letters. In the pattern databases for each word, for each group of characters and for each letter can be put into correspondence the individual pattern surface.

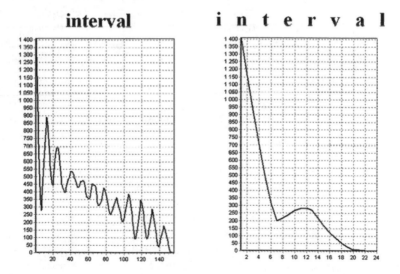

Fig. 5.11: $FAI(x)$ of the word 'interval' at different distances between the letters

Usually people during analysis of the visual scene pay attention only to some of the main objects; other information about the environment is processed as some model. The system built of parallel shift technology allows simulation human vision. However, for accurate modelling it is necessary to build hybrid systems of technical view. They should include the possibility of digital image processing. The hybrid human vision can be explained as follows. In determination of the movement parameters, taking into account the Doppler effect, the object moves much faster than the shift copies. This movement will not be registered, as if there was no object in the visual scene. This is similar to the effect of 25 frame, but this effect confirms the presence of other information-processing systems, where the entire video stream is divided into frames, stored and processed subconsciously.

5.4 Perspective Directions of Image-processing Research Using PST

The system of images perception, which is used in the construction of parallel shift technology, has advantages in performing certain image-processing tasks. In the process of implementation of the algorithm, the information is aggregated and ordered that, in some cases, it leads to productivity improvement, but sometimes, these properties make it impossible to perform certain tasks.

Let us consider the image-restoration issues processed by using the parallel shift technology. If image description is only a set of functions of the area of intersection, then restoration is possible in the following ways.

Suppose we know the initial image area (parameter S_0). Let us divide this value into the largest possible number (n) of identical elements. The area of each of them is equal to $\frac{S_0}{n}$. The visual field is divided into sections of the same area. The complete enumeration method determines all the possible options for the location of n elements of the object in the visual field. Thus, it is possible to choose the location of the image elements, such that in the formation of set FAI, these features are close in value to the elements of the pattern surface, which describes the object. The process of performing the complete enumeration is very simple, but requires a very large amount of time to perform.

The second method may be stored in the image-copy memory. If required to restore the data from memory, the stored image is scaled to the required size. The FAI set in restoration is not involved. In this case, additional costs of storing reference image are needed. Also, if a certain class of images is stored in one pattern, then in restoring it cannot be identical to the required image.

To reduce the stored information, the pattern can be represented by a set of syntax expression for simple shapes. Then the image reconstruction will be quite a laborious process. Accuracy of restoration can also be rather crude.

Thus, the image-restoration process using vision systems based on PST is not productive. In contrast, the restoration of digital image is an elementary process.

For the analysis of scenes, two sub-tasks (Duda and Hart, 1973) can be selected. The first is the search for the specified image as part of a single object in the visual scene. The second is the search for specified images among the many individual elements of the scene. An example of these tasks is shown in Fig. 5.12.

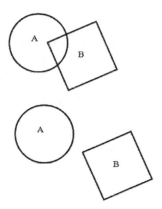

Fig. 5.12: Various tasks of identification and location of the image

In the first case, it is determined whether there are A and B elements in the visual scene. The scene consists of a single object AB.

$$S_{AB} = S_A + S_B - S_{A \cap B} \neq S_A + S_B. \tag{5.5}$$

In the second case, it is necessary to determine the presence and localisation of A and B elements. But in this case, the scene consists of elements, which can be considered as a separate entity, as well as one whole unit.

$$S_{AB} = S_A + S_B. \tag{5.6}$$

In determining the FAI of the composite figure the following formula should be used

$$FAI_{(x)AB} = FAI(x)_A + FAI(x)_B + FAI(x)_{A,B} - FAI(x)_{A \cap B}. \tag{5.7}$$

where $FAI(x)_{A,B}$ is the area of intersection of the copy of A figure with the B figure. The $FAI(x)_{A \cap B}$ component in the first case occurs when initially the A figure intersects the B figure. In the second case, the component is equal to zero.

These examples are presented with two figures which are present in the visual scene. For three figures, the formula (5.7) takes the following form:

$$FAI_{(x)ABC} = FAI(x)_A + FAI(x)_B + FAI(x)_C + FAI(x)_{A,B} + FAI(x)_{A,C}$$
$$+ FAI(x)_{B,C} - FAI(x)_{A \cap B} - FAI(x)_{A \cap C} - FAI(x)_{B \cap C}. \quad (5.8)$$

Formulas are presented for the shift copy direction x; for other directions of FAI formation, the formulas are similar.

Each function of the area of intersection can be represented as a set of harmonics of Fourier series. Then you can analyse the FAI spectrum of the entire image and the function of the area of intersection of each figure separately.

Functions of the area of intersection depend on the interaction of the image elements at a certain time. Therefore, their analysis can be carried out similarly to the analysis of transient processes in electrical networks. For example, the image can be associated with an electrical circuit and the function of the area of intersection can be considered as the integral of Duhamel. Thus, using the parallel shift technology the process of object conversion in the function can be organized and vice versa. This task includes image reconstruction subtasks and scene analysis. Research in the given direction has not yet been carried out, but according to the authors, the prospects are promising.

One of the elements that significantly affects the manipulation of the image in the process of its processing which is need to carry out FAI scaling. The need for scaling appears during comparison of functions. This is due to the fact that the pattern FAI by parameters can differ from the functions of the area of intersection of the real image. The solution to this problem may be a possibility of object representation not in a surface that is constructed based on the FAI set and as the integrated coefficient set for the respective directions. In this case the scaling is not needed during the comparison. To apply this form of object representation it is necessary to organise a parallel shift of its copy to all directions of 0 to π.

Thus, to describe the object the vector of the values of the integral coefficients can be used. The dimension of this vector can be limited by forming an approximate pattern surface. In the process of computer simulation of image recognition process, the amount of shift in directions equals 36. Presentation of the functions of the area of intersection as a harmonic of a Fourier series can also be limited to a certain number. In the process of computer simulation of image-recognition process, the number of harmonics of the Fourier series was 20 for a description of the function of the area of intersection.

In both cases we have a small number of homogeneous elements to describe the objects. Each of these vectors can be processed using neural networks (Milanova et al., 1999; Ranganath and Arun, 1997; Lawrence et al., 1997; Yoon et al., 1998; Aizenberg et al., 1999; Minsky and Papert, 1969; Kozhemyako et al., 2001). A small amount of input data allows use of a small number of neurons for its construction. The values of the integral ratios do not exceed 0.5. Values of normalised harmonics of the Fourier series are close to unity. These values make it easy to organise the vector of input information to the neural network.

The computer-image recognition processing system based on an analysis of indicators of the function of the area of intersection is designed to handle any flat image, and can be used in any video surveillance system. Since the process of the function of the area of intersection gets involved in processing large arrays of video, then to accelerate the process it is expedient to build high-performance specialised devices.

6

Assessment of Productivity of the Image Processing and Recognition Systems Based on FAI Analysis

The productivity of the image processing system may be evaluated according to various parameters. These indicators include the amount of information that is processed per unit of time, or the percentage of correctly recognised images among the plurality of images.

Determination of the speed of information processing for image recognition system based on the analysis of the functions of the area of intersection is not possible. At the time of writing, the hardware that constitutes such a system had not been built. The software modelling was performed using the Delphi 7 programming language. In the programming model of image recognition system a working field has a size 500×500 pixels. Its square shape is necessary for implementation of the process of rotation of the object in the learning phase. In this case, on the formation of a single FAI at an average of one second is spent. Functions of the area of the intersection of real image are determined for two of the orthogonal directions; also, the time spent directly in comparison of the FAI and integral coefficients of the real image with the image patterns. Obviously, for hardware implementation due to parallel processes the time that is necessary for processing of objects will be less by two orders.

In future, an assessment of productivity of the image processing system based on analysis of the FAI will be carried out by the criterion of the quality of recognition. These processes simulate preliminary stages of image processing and analysis approach to the real image. Therefore, in this case, it is not possible to conduct identification without using acceptable parameters of errors. The image recognition system, based on the analysis of function of the area of intersection, uses three types of permissible errors.

Error dk is used when comparing the integral ratios of the function of the area of intersection in the fast stage of the recognition process; error dS is used for comparison of the functions of the area of intersection at a detailed stage of the recognition process and error dN is used in the formula of value determination of maximum shifts.

Use of these errors in the image processing system is necessary especially in connection with certain rounding off for mathematical calculations, loss of information when converting data arrays and for the possible presence of noise in the receptor field.

6.1 Evaluating the Efficiency of the Chosen Method of Removing Noise

An important factor for each method of data processing is the selection of the removal of noise algorithm. The areas of the figures are analysed by methods of preprocessing and of image recognition on the basis of parallel shift technology. Noise interference in this case is considered as the individual element of the local groups and the contour images and lines. In addition, the noise will be considered as null because elements of the same nature can occur in place of the binary elements constituting the binary image.

To fight with noise the contour points removal method, which has been described previously, may be employed. This method is quite effective at values of PSNR of 20 dB to 40 dB. In this range of the noise level, the deviation of scale values of the FAI and a variety of additional factors do not exceed 10 per cent of the value of real indicators.

However, the removing point contour method has been in design stage during the creation of software model of the provisional processing and image recognition process. Therefore, to simulate the phase of noise control, 3×3 square filter with a threshold of 5 is used. When the sum of the binary elements of the matrix 3×3 is greater or equal to 5, the central element is assigned a value of 1; otherwise it is set to 0. As shown below, the process of noise control while using this filter is effective for values of PSNR of 20 dB to 40 dB.

It is also possible to use for noise control of 5×5 square filter with the threshold value 12. For all the three ways of dealing with noise, research on the effectiveness of work were carried out and the research results will be displayed further.

The value of the amount of noise elements among all the elements of the image can be determined, by using the value of noise inclusion percentages (pN). The values of this indicator are in the range 0 to 100 per cent. A widely used PSNR parameter is a result of logarithmic dependence. This type of dependence is used in a wide range of input data. When input data differ from each other by not more than two orders of magnitude, it is preferable to use a simple parameter of the noise percentage. We will show the relationship of the parameters *PSNR* and *pN*.

$$PSNR = 20\lg\left(\frac{MAXi}{\sqrt{MSE}}\right),\qquad(6.1)$$

where *MAXi* is this maximum value received by the pixel image. For a binary image *MAXi* = 1.

The mean square error of the MSE, expressed in terms of the percentage of noise, will be as follows:

$$MSE = \left(\frac{pN}{100}\right)^2.\qquad(6.2)$$

These formulas allow us to determine the relationship of the parameters *PSNR* and *pN*.

$$PSNR = 20\lg\left(\frac{100}{pN}\right).\qquad(6.3)$$

Some values of conformity of these parameters are shown in Table 6.1.

Table 6.1: Some parameter values PSNR and pN

pN, per cent	PSNR, dB
1	40,00
5	26,02
10	20,00
20	13,98
30	10,46
40	7,96
50	6,02
60	4,44
70	3,10
80	1,94
90	0,92
100	0,00

We define an algorithm of filling the receptor field by the noise elements.

Let there be a matrix A of size $i \times j$, which consists of elements $a_{i,j}$, where $a_{i,j} \in [0; 1]$. Then, if $a_{i,j} = 0$ and the value of random function (100) = 1, $a_{i,j} = 1$. If $a_{i,j} = 1$, the value of random function (100) = 1, $a_{i,j} = 0$. In this case, the noise components have inverted value of the input binary image. The input image matrix is filled approximately by one per cent of the noise components. The number of possible noise per cent (pN) for each element is the probability that it is noise. This procedure must be repeated pN times to achieve the desired result. Thus, $pN = 100$ per cent is filled by the unit cells of half pixels of the binary matrix, which consists of zero elements. This occurs due to the fact that in the process of inversion there can be nulled elements, which previously were assigned a 1 value. However, this property does not affect the image recognition process for those values of noise where the noise removal methods are effective.

Thus, it will be obtained for the number of noise image elements kN, depending on the percentage of noise.

$$kN = S_A \cdot \frac{pN}{100}. \tag{6.4}$$

S_A is the total area of the receptor field. Since in the programming model its dimensions were taken as 500×500 pixels, then $S_A = 250000$.

An indicator of the efficiency of the method of removing noise will have a certain amount of noise components (kNc), that have remained in the receptor field due to implementation of noise removal process. This amount should be such that its influence in the analysis of the FAI will be offset by the use of permissible error dN. It should be noted that the efficiency of noise removal process in this research is the ability to delete the 1th element into initially empty matrix.

To analyse efficiency of the given method, the amount of noise elements kN determined and the number of non-deleted noise elements kNc was carried out for all the possible noise per cent pN, where $pN \in$ [1..100]. The studies were conducted for each of the three specified methods for combating noise. The procedure of this research is as follows:

1. Each value of the noise per cent produced is filled with the empty image by noise interference.
2. The amount of noise elements is calculated as $kN = \sum\limits_{n=1}^{i} \sum\limits_{m=1}^{j} a_{n,m}$.
3. The noise removal procedure is performed.
4. The amount of the remaining noise elements is calculated as

$$kNc = \sum\limits_{n=1}^{i} \sum\limits_{m=1}^{j} a_{n,m}.$$

The determination of hundred kN and kNc for hundred pN was carried out.

For each i-th pN_i value of average values kN_i and kNc_i is determined. In order to generalise the data, it is necessary to carry out the normalisation of the obtained mean values of the number of noise components (kN and kNc) by dividing into the value of the total area of the receptor field. Thus, the indicators independent of the specific predetermined area will be obtained S_A.

$$kN_i = \frac{\sum\limits_{j=1}^{100} kN_j}{100} \tag{6.5}$$

$$kNc_i = \frac{\sum\limits_{j=1}^{100} kNc_j}{100} \tag{6.6}$$

Value of the coefficient of efficiency is calculated by the following formula:

$$eN = \frac{kN - kNc}{kN}. \tag{6.7}$$

Generalised data of research on the effectiveness of the noise removal process is shown in Fig. 6.1.

As seen in Fig. 6.1, the efficiency factor (eN) of each noise removal method varies depending on the percentage of noise interference pN.

When using the matrix 3×3 to combat noise, the following results will be obtained. As shown in the chart, four ranges of noise values can be selected. The first is the kNc value which is equal to zero ($pN \in [0..6]$). The method of noise removal is most effective ($eN = 1$). Noise items are fully removed in the process of noise control. Second is the kNc value which is close to zero ($pN \in [7..18]$). The method of noise removal is most effective ($eN \approx 1$). Here the noise has almost no effect on recognition, but the identification process requires additional adjustment through the use of the error dN. The third is a small kNc value ($pN \in [19..30]$). The method of noise removal is effective ($eN > 0,85$) but the recognition process is complicated. There are a large number of errors in identification. Fourth is a sharp increase in the value kNc ($pN \in [31..100]$). A noise removal method is not effective and recognition is not possible.

When used for noise control of the matrix which is 5×5 size, the following results are obtained. As shown in the chart, the four ranges of noise values can be selected. The first is the kNc value which is equal to zero ($pN \in [0..9]$). The method of noise removal is most effective ($eN = 1$). Noise items are fully being removed in the process of noise control. The second is the kNc value which is small ($pN \in [10..26]$). The method of noise removal is effective ($eN > 0.90$), but the recognition process is difficult. There are a large number of errors in the identification process. The third shows a sharp increase in the value kNc ($pN \in [27..53]$). A noise removal method is not effective and recognition is not possible. Fourth is the efficiency ratio which is less than zero ($pN \in [54..100]$). The method of noise removal adds the noise elements in the receptor field. A noise removal method is not effective and neither is recognition not possible.

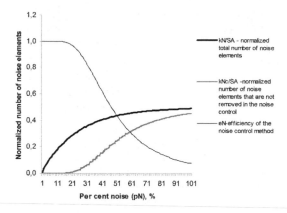

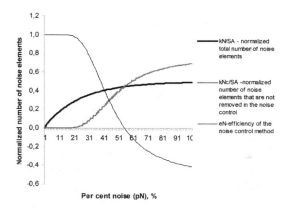

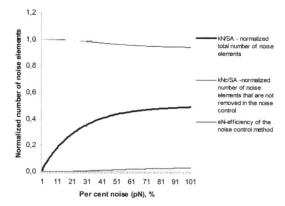

Fig. 6.1: Generalised data of research on the effectiveness of noise removal
process by different ways

For removing the contour points method, the effectiveness coefficient calculated by the formula (6.7) will be close to one for the entire range of values pN. But unlike the methods of matrices application with the threshold value, also the method removes the contour points of the image. Large percent values of noise inclusion in noise control must be removed because too much useful information distorts the original object. Recognition becomes impossible. This method is effective for values $pN \in [0..10]$.

The use of one or other method of struggle against noise does not affect the principles of image processing using parallel shift technology. Investigation of these methods to calculate the values of permissible errors parameters is necessary, which, in the analysis of the FAI is used. In future, the calculation of the values of these parameters and efficiency of object recognition will be carried out using noise control of matrix 3×3.

6.2 Determination of the Permissible Error Range of Values Used to Limit the Maximum Displacement Values

The possible presence of noise impurities in the receptor field during recognition of real images encourages the introduction of certain adjustments in the calculation process. Limitations of values of parameters of the maximum shifts is produced by application of the permissible error dN. We calculate the dN parameter depending on the number of kNc image elements that are not removed in the process of noise control. The average value of noise elements in number of real images, which can get into the shift region of the copy at each time point, is denoted as kNc_0. With a uniform distribution of noise, this parameter has the following value:

$$kNc_0 = kNc \cdot \frac{S_0}{S_A} \tag{6.8}$$

Image recognition is possible only if the value is less than the limiting values $dN \cdot S_0$, used in calculation of the maximum shear parameter. The use of this property is the third level of noise control. Thus, the dN value will be as follows:

$$dN \geq \frac{kNc}{S_A} \tag{6.9}$$

Hence, the lower boundary of the dN values for each noise level corresponds to the normalised number of noise elements and are not deleted in the fight against noise as seen in Fig. 6.1a. So, when 7 per cent $< pN < 18$ per cent, the value $dN < 0.0048$. Thus, for noise levels with $pN < 18$ per cent, the upper boundary of the range of the permissible error dN

values is determined. With a small amount of noise, the $dN < 0,005$ value can be used.

Since for 19 per cent $< pN <$ 30 per cent, the value $dN < 0.0634$, but at higher pN values the noise removal algorithm is not effective. The upper boundary of the range values of the permissible error dN should be close to 0.07. These graphics confirm these values when used for noise control of the matrix 5×5. With software modelling of the processes of image preprocessing and image recognition based on PST the dN range from 0 to 0.2 is used.

6.3 Determination of Values Range of Acceptable Error Used in Comparison of Integral Coefficients

In the process of comparing the integral ratios on the stage of image recognition by the method of analysis of the functions of the area of intersection, the dk value of permissible error is used. The need to use this parameter is determined by the fact that in calculating the basic parameters of the function of the area of intersection or in the construction of their basic functions for simple shapes, certain data rounding off is done. It leads to the fact that the expected data are different from the real values. For all types of recognised images, the noise removal process is used. This fact also influences further calculations because some parts of useful information can be deleted. In addition, data losses occur as a result of processing the input information arrays.

The basis for determining the amount of allowable error dk is in the range of the integral coefficient values. The integral coefficients (k_φ) take the values from 0 to 0.5. Furthermore, efficient use of these parameters is caused by the opportunity to separate the general-class patterns (A) into several subclasses (A_i), in order to restrict the number of patterns for further detailed comparison of the function of the area of intersection. The number of subclasses (nA_i) is determined from the following formula:

$$nA_i = \frac{0,5}{2 \cdot dk} = \frac{1}{4 \cdot dk}. \tag{6.10}$$

Separation of the pattern into subclasses is quite conditional and depends on the integral values of the coefficients k_φ. Graphically the selection of ranges of k_φ values is reflected in Fig. 6.2.

For maximum efficiency of the integral factors in the image recognition process, the number of subclasses nA_i should be maximally large. Accordingly, the dk value should approach zero. In this case, for a detailed comparison of the functions, the smallest number of patterns is transferred. According to (6.10) the value of allowable error dk cannot exceed 0.25. This statement is valid where there is only one class of

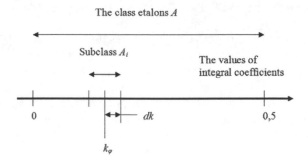

Fig. 6.2: Formation of ranges of subclasses of integral coefficients based on permissible accuracy *dk*

templates ($nA_i = 1$). With increase in the number of subclasses nA_i the required value of this parameter will only decrease.

With technical implementation of the pattern recognition system based on PST, the calculation roundings, loss of data and other information distortion are possible. These factors limit the values of this error. Due to the mentioned circumstances, the formula for real integral factors will change slightly.

$$k_{xN} = \frac{kNc \cdot \dfrac{S_0}{S_A} \cdot X_{max} + \int\limits_{0}^{X_{max}} FAI(x)dx}{(S_0 + kNc)X_{max}} \tag{6.11}$$

where k_{xN} is the real integral coefficient for the x direction in the presence of noise in the image, kNc is the amount of noise elements that are not deleted and S_A is the total area of the receptor field.

Then the value is

$$dk \geq |k_{xN} - k_x|. \tag{6.12}$$

After transformation, this formula takes the following form:

$$dk \geq \frac{kNc}{S_A} \left| \frac{S_A - S_0}{2S_0 + \dfrac{kNc}{S_A}S_A} \right|. \tag{6.13}$$

The data of normalised values of the number of elements that are not removed in the process of noise control using 3×3 matrix can be taken from Fig. 6.1a. The ratio of area of the receptor field to the area of the image will be called rS.

$$rS = \frac{S_A}{S_0}. \tag{6.14}$$

The graphic form of the permissible error dk values depending on the rS parameter values when $pN = 17$ per cent is shown in Fig. 6.3.

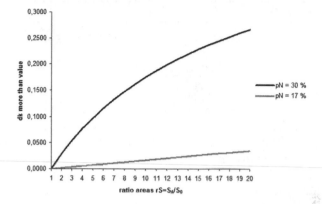

Fig. 6.3: The possible dk values depending on the rS parameter

In the process of software simulation for testing the dk values ranging from 0 to 0.04 is used and which correspond to the calculated values for $pN < 30$ and $rS \leq 4$. Value at rS calculations is selected for organisation of cyclic shift in two orthogonal directions and the object dimensions in each of them shall be no more than half the size of the receptor field.

6.4 Determination of Range Values of Acceptable Error in Comparison of FAI on a Detailed Stage of the Analysis of Image Recognition

On the detailed stage of pattern recognition by the method of analysing the functions of the area of intersection, the acceptable error dS is used. At this stage, a comparison of the actual results and standard functions crossing areas with a given accuracy dS is performed. The necessity to use this parameter is caused by similar circumstances that require the application of allowable dk error.

For the lower bound of the range of possible values, this parameter depends on the changes that appear in the image due to the influence of the noise removal process.

$$dS > \frac{|FAI(x)_N - FAI(x)|}{S_0},$$

(6.15)

where $FAI(x)_N$ is the value of the function of the area of intersection of image, which was corrected by noise removal element process.

The calculated data of the allowable errors dS for simple shapes with $pN = 30$ per cent in Table 6.2 are shown. The data are based on the maximum values of the rS parameter, which was used in software simulation.

Table 6.2: The calculated data of the allowable errors dS values for
simple shapes with $pN = 30$ per cent

Name of the simple figure	rS	dS
Triangle	≈ 10	$\approx 0,16$
Circle	≈ 5	$\approx 0,1$
Square/rectangle	≈ 4	$\approx 0,08$

From the data table it follows that the value of the acceptable margin of error, which is used in the comparison functions of the area of intersection of the simple shapes with $pN = 30$ per cent, must be greater than 0.16. However, at lower noise levels, a given value may be smaller. In the software simulation process for testing, the dS value range from 0 to 0.8 is used for the possibility to investigate the recognition efficiency depending on the value of this parameter.

Values of permissible errors were calculated to approximate initial evaluation of these parameters when building software models of the image recognition processes through methods based on the parallel shift technology. Various studies on the efficiency of these methods have been carried out, using a software simulation.

6.5 Evaluating the Effectiveness of Image Recognition Process in Various Parameters of Input Information Using Calculated Values of Permissible Errors

To evaluate the effectiveness of the image recognition process with different parameters of input information, it is necessary to define the type of errors, the occurrence of which is possible in this process.

As the first kind of mistake, we shall consider the case when the figure is in the database of templates, but in the process of identification, is assigned to the class of images, consisting of more than one element, and in this class is included the figure that is being recognised. A typical example of an error of this kind is optional, when the parameters of permissible errors are selected incorrectly.

The second kind of error will be called in a situation when the figure is in the template database, but on the detailed step of identification it is assigned to the image class, which does not contain the given image. The main cause of error appearance of this kind is a high noise level. The negative influence of this noise cannot be deleted by the chosen method of noise removal.

The third kind of error occurs when there is a figure-based template, but is not found in the database.

On the basis of the calculated parameter of permissible errors, recognition of simple shapes was carried out to determine the effectiveness of the process according to the existing per cent of noise elements in the input image (*pN*). The following parameters were used to determine the coefficient of efficiency:

kC – number of circle images provided for recognition.

kCn – number of circle images that are not recognised by the system.

eC – efficiency coefficient of circle recognition.

The efficiency coefficient value of circle recognition is determined by the following formula:

$$eC = \frac{kC - kCn}{kC}. \tag{6.16}$$

We denote the analogous parameters for the square (*kS, kSn, eS*), for the rectangle (*kR, kRn, eR*), for the triangle (*kT, kTn, eT*).

In the process of figure recognition, the following errors are used *dk* = 0.04, *dS* = 0.1, *dN* = 0.01. These values correspond to the analytically calculated range.

The process of identification of the figures was carried out to find the noise inclusion per cent from 0 to 20 corresponding to a range of effective noise removal by the chosen method. A total of 500 test images was submitted for each class of simple figures for each *pN* (*kC* = *kS* = *kR* = *kT* = 500). Using the obtained coefficients values, the average value of the coefficient of efficiency of recognition of simple shapes (*eAs*) was calculated for the *pN* parameter equaling 0 per cent, 5 per cent, 10 per cent, 15 per cent and 20 per cent. Efficiency coefficients are calculated by the following formula:

$$eAs = \frac{eC + eS + eR + eT}{4}. \tag{6.17}$$

Graphically, the average value of the efficiency coefficient of simple shapes recognition depend on the percentage of the noise is shown in Fig. 6.4.

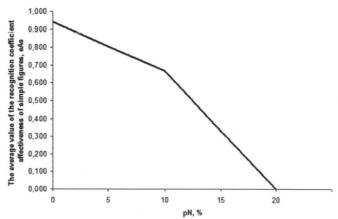

Fig. 6.4: Average values of efficiency coefficients of simple shapes recognition depending graphically on the noise percentage

The results of these researches show a correlation of efficiency of application of parallel shift technology to recognise simple shapes with efficiency of the chosen method of noise control. Furthermore, it should be noted that the data values affected by the input image area were arbitrary. Generation of small objects have a negative impact on the effectiveness of the identification. The relationship between the area of the input image and the recognition performance will be considered in future.

For standard image recognition process it is impossible to analytically determine the error values. In the process of recognition of figures, the following error values are used $dk = 0.04$, $dS = 0.1$, $dN = 0.01$. These parameters of the permissible errors are similar to those used in the investigation of simple shapes.

For determining the efficiency of the recognition of pattern shapes in an array of patterns was adopted by a set of numbers and letters of Latin and Cyrillic alphabets printed in Arial font with size 72 pt. It should be noted that this system is not specialised for recognition of the printing font. This type of standards was selected for ease of creating template images and due to the variety of their forms (Krzyzak et al., 1990).

To investigate the effectiveness of the identification of template images based on the percentage of noise interference in the recognition system, 700 images of numbers and capital Latin letters of Arial 72 pt font were submitted. Test images were submitted for each value of the interference noise per cent (pN) from 0 per cent to 20 per cent with a step of 5 per cent. Accordingly, each of the images was processed 20 times for each noise per cent. The only exception was the letter 'I' as in Arial fonts it is a rectangle, recognised by the system as corresponding to a simple figure. This fact makes it impossible to use this image for assessing the effectiveness of recognition of standard shapes. Graphic display of efficiency recognition ratio of the pattern figures (eE) depending on the percentage of the noise is presented in Fig. 6.5.

$$eE = \frac{nE}{700} \tag{6.18}$$

where nE is the number of correctly recognised patterns for each noise per cent.

There is an obvious correlation between the effectiveness in recognition of simple shapes and template. Except for the presence of noise inclusion on the efficiency of image recognition, the use of TPS is influenced by other factors.

The presence of the first kind error indicates that the fast identification step for the images is passed successfully. Non-uniqueness and the definition of standard class of shapes into which the test objects are added, occurs due to the following reasons:

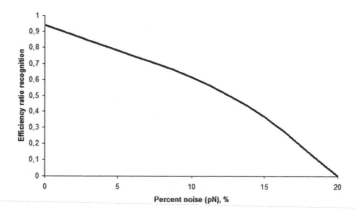

Fig. 6.5: Graphic representation of values in effective coefficients recognition of etalon figures, depending on the percentage of noise

1. The accuracy of the restoration of the function of the area of intersection of pattern values when using fast Fourier transformation does not match the recognition dS value.
2. The system cannot distinguish between the mirror shapes (e.g., letters R and Я, P and b).
3. Images of some of the Arial symbols in orthogonal directions provide the same FAI (e.g., letters H and П).
4. Some images for the chosen values of allowable errors (dk and dS) give a similar result; i.e., the system depending on invariance to the angle of rotation can identify certain images as one (e.g. Figs 6 and 9).

Reasons for decision of ambiguous recognition are the following. Accuracy of the FAI restoration can be enhanced by storing patterns in the form of a greater number of fast Fourier transform harmonics or of direct conservation of the patterns as a set of functions. Both ways envisage an increase in the pattern database with special attention on the processes of the scale functions for comparison at the detail stage. To recognise the mirror images, or to solve the problem of the third cause of appearance of the error of the first kind, it is necessary to perform a second scan of image in a direction that is different from the first, not on 90°, but on the other corner. The FAI formation may also be performed for three (or more) shift directions that will identify an image using a tuple of three (or more) elements. With the help of such methods, the influence of the central symmetry of the template surfaces can be neutralised. Such a representation of an object requires a high accuracy in calculation of the process organisation of non-orthogonal parallel shifts.

It should be noted that the letters have less convex shapes than the simple images. Therefore, the efficiency of this method of pattern recognition using parallel shift technology depends on the degree of convexity of the object, which is recognisable. This fact is a consequence of the changes in the results of a detailed process of the area of intersection comparison due to loss of information during stages of scaling and transformation functions.

One of the main parameters used for image recognition based on the use of parallel shift technology is the area of the input shape (S_0). There is a necessity in research of the image recognition efficiency, depending on this parameter.

The digitised input image is a matrix of elements that is created by squares with size 1×1. Therefore, for small values of input image area, will be significantly changed at the step of perception, resulting to distortion of information. For example, a circle with a diameter of 2 pixels is represented as a square 2×2. Furthermore, in the presence of noise elements or during the deleting of noise, the initial area can be significantly changed. It will affect the process of identification. It is therefore important to determine the minimum size of the border of the input image in which further recognition is possible.

To determine the efficiency of image recognition depending on the value of the area of the input figure, the statistical data collection was done. Several groups of 2000 simple shapes with parameters of allowable errors ($dk = 0.04$, $dS = 0.1$, $dN = 0.01$) were selected with a percentage of noise inclusions (pN) 0 per cent and 10 per cent. The data obtained from this process is reflected in Fig. 6.6.

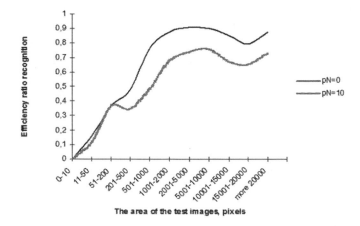

Fig. 6.6: Graphic representation of the values of effective coefficients recognition of simple figures, depending on the area of test images

It can be concluded that if the initial value of the image area is in the range of 2000 pixels or more and the percentage value inclusion noise is

0 to 10, the recognition efficiency ratio is greater than 70 per cent. This should be borne in mind that any percentage of image noise interference at the start of the process of removing noise is carried out. This somewhat impairs the efficiency of identification. Local extremes of functions are caused by disadvantages in the generation of test images with software simulation. However, this fact does not change the overall picture of the efficiency of research in image recognition according to the initial area.

It is necessary to substantiate that the assertion of invariance in the recognition method is relative to the size of the template figures. For this purpose, the coefficient of identification efficiency of the template images of various sizes is investigated.

The test images have selected numerals and capital letters of the Latin alphabet in Arial font of sizes 72, 80, 90 and 98 pt. These sizes of the font symbols have an area that is close to the area of the effective recognition of simple shapes. Furthermore, the coefficients ($k\rho_0$) of the density of arrangement of elements have almost a similar size of fonts. Hence, the same symbols of those fonts have the same shape. This restriction is imposed because the same symbols of bitmap fonts can have different shapes because the values of their linear dimensions are integers. They have a hopping form. A graphic representation of the efficiency coefficient values of the recognition template shapes, depending on their initial area is shown in Figure 6.7.

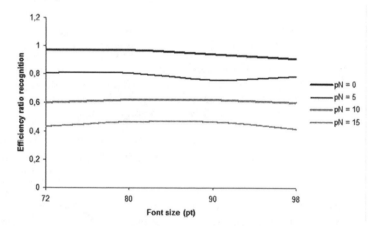

Fig. 6.7: Graphic representation of effectiveness values of coefficients of etalon figures, depending on the area of test images

Based on the results of this study, one can conclude that image recognition based on parallel shift technology is invariant with respect to the size of the input objects. Besides, the efficiency in identification of different size shapes depending on the amount of noise interference is within the values obtained in previous studies.

Storing a template as image database of values for different scanning directions allows independence in the recognition process not only on the image scale, but also on its orientation direction. Let us consider a study to confirm the invariance of the proposed method to such an affine transformation as a rotation. For this purpose, we choose an array of test objects. They are the numbers and capital Latin letters in Arial font size of 72 pt. Each of the pattern images is described by using a template surface. Each surface is composed of 36 functions of the area of intersection in directions from 0 to 175 ° in steps of 5 degrees. This study was carried out for values of noise interference percentages of 0, 5, 10 and 15. The values of allowable errors that are selected are same as in the previous study. Data of research of test image of recognition efficiency are shown in Fig. 6.8 (Foltyniewicz, 1995).

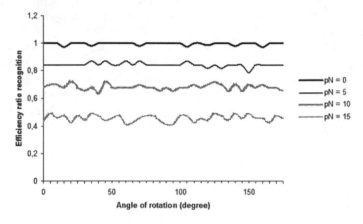

Fig. 6.8: Graphic representation of effectiveness values of the recognition coefficients of etalon figures, depending on the angle of rotation

It should be noted that the effectiveness of image identification negatively affect the bending shapes of objects in software implementation on a rotation stage training system and in the course of research (Jacobsen et al., 1999).

Data of research in image recognition process given in this book show lower values effectiveness than in previous publications. This happens because in this study a much more sophisticated type of noise was used. Visually estimation of the two types of noise is seen in Fig. 3.5.

Investigations that were carried out by means of the software model confirm the invariance of the proposed recognition methods in the sizes and angle of rotation of the original image.

Hardware Implementation of the Image Recognition System Based on Parallel Shift Technology

7.1 Generalised Model of Image Recognition System Based on Parallel Shift Technology

An important factor in the theory of image recognition based on parallel shift technology is the hardware implementation of the proposed methods, which has good characteristics in comparison with existing methods. The hardware costs for implementing the proposed methods should be minimal and simple, having high performance with minimal power consumption.

The feasible variants in realisation of the recognition system based on the technology of parallel shift is presented in Fig. 7.1.

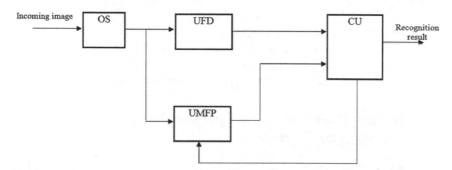

Fig. 7.1: Block diagram of the image recognition device based on the parallel shift technology.

The image recognition system comprises an optical system (OS), the function of the area of intersection detection block (UFD), the memory block of the function of the area of intersecting etalons (UMFP) and a comparing unit (CU).

The input image through the OS is converted into the required format to enter the UFD. The UFD calculates the function of the area of intersection for the shift directions that are set by the user. The quantitative values of the function of the area of intersection are fed to the first input of the CU in comparison to all the template functions of the area of intersection in the UMFP. If the function of the area of intersection at the first input of the CU coincides with one of the template functions of the area of intersection from the UMFP at the second input, then the identifier of the template FAI is generated at the output of CU. This identifier indicates the membership class of the input image.

If the UMFP does not have the function of the area of intersection of equal FAIs at the first input of CU, the input FAI is recorded in the UMFP and is assigned an identifier of belonging to the class.

In Fig. 7.1 a well-known scheme of the recognition system is presented (Bilan, 2014). It does not have a detailed description of a specific method nor does it describe how the UFD determines the FAI in several directions and how the copy of the input image is shifted. In addition, neither is the FAI value represented nor a description of the function of the area of intersection representation and its storage in the UMFP are given.

If we use the function of the area of intersection as a whole, then it can be represented by a set of values of a set of pairs of number $\{X_i, S_i^j (x)\}$, (where i is the step number of the copy image shift in the j direction). To store all the functions of the area of intersection of a single image, $i \times j$ memory cells are necessary. Also, additional memory is used to specify the class and subclass identifiers. If a CU is used with a sequential structure, the input for one FAI can be k bits that are used to represent a single value of X or S. Thus, for comparison of one of the functions of the area of intersection, i time steps are used. This value is determined by the number of shift steps in one direction.

A parallel comparison of the entire FAI code uses a parallel reading of the template functions of the area of intersection from the UMFP. Both implementations can be used for certain characteristics of the system.

7.2 Selecting a Cover for Representing the Image in the Recognition System

With such a device structure, the shifts for each selected direction are realised sequentially. Initially, the right shift is performed, then upwards, and so on. The more the shift directions the more accurate is the description of the image at the input of the system. However, the accuracy in calculating the area of intersection depends on the discretisation of the field, which displays the original real image in the system. In the example (Fig. 7.2) an orthogonal covering is used. This coating makes it possible to determine the function of the area of intersection without losses in four directions (right, left, up, down), since there is no distortion of images during the shift.

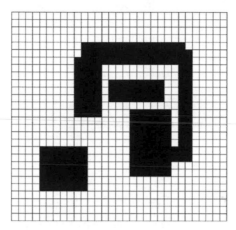

Fig. 7.2: An example of image forming on the orthogonal cover

Orthographic coverage gives it a unique description for most images that distinguish it from other images. However, there are cases of complex symmetric images for which the function of the area of intersection in four directions can coincide. An increase in the number of directions with an orthogonal covering does not give an accurate representation of the intersection area. To eliminate this disadvantage, other forms of coating are used. The most effective and popular form of coverage is shown in Fig. 7.3 (Belan and Motornyuk, 2013; Nicoladie, 2014; Bilan et al., 2014, Konstantinos, 2011).

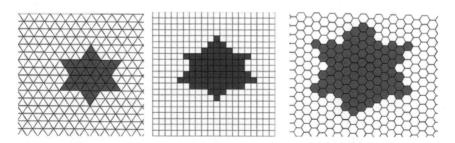

Fig. 7.3: The most common form of coverage and discretisation of images

All these forms of coverage accurately represent a real analog image. However, to realise the image shift in a large number of directions, the hexagonal form of the coating is most effective as it gives six directions of image shift without distortion. Even a hexagonal mosaic can give twelve shift directions, which are determined additionally on the vertices without loss of information. In this case, the additional six shift directions are formed along the vertices and give a larger step of discretisation of the shift by the value of one side of the cell. These directions are given in Fig. 7.4. The same can be said for other forms of coating. However, for hexagonal coverage, increase in the number of directions gives less distortion of the original image.

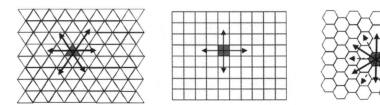

Fig. 7.4: Directions of image shift without distortion for triangular, orthogonal and hexagonal forms of coating.

If the software model for implementation of the shift is used, then the coordinates of the image points for each step of the shift are calculated again. However, a discrete field cannot always give an exact result of a shift for directions at small angles.

Hardware implementation allows implementation of an image shift without first calculating the coordinates of the image pixel shift. Analysis of the examined coatings showed that there are only three types of lattices that are built on the basis of regular polygons that makes it possible to form a dense mosaic field. Such regular polygons include triangles, quadrilaterals and hexagons. Use of other grids of regular polygons does not make it possible to form a discrete plane densely. Other forms of coverage with different forms of cells do not allow the shift of images without loss of information.

7.3 Hardware Implementation of Parallel Shift and Computation of the Area of Intersection of Images

Consider an image recognition system with a detailed structure of the FAI detection units and other devices based on them. Let's look at the detailed implementation of each block of the system.

The first unit that begins to function is OS. With the help of this system, the necessary transformation and preparation of the image for calculating the function of the area of intersection are implemented. We will not consider this system in the work, since it is not part of the implementation of the proposed methods and can be selected from known implementations of such systems.

The main value in the implementation of the method is UFD. It is this unit that computes the function of the area of intersection and presents it in an appropriate form.

The first operation that the block should implement is to generate a copy of the projected image. For this, a two-layer matrix medium is used. Each layer is represented by a matrix. The first and second matrices store the initial image. In the first matrix, the image is fixed and the second matrix shifts the image in the selected direction. The intersection area is determined by an additional layer. Each cell of this layer gives signals if it belongs to the intersection area.

The structure of the device for calculating the area of intersection and isolation of cells that belong to it is shown in Fig. 7.5. The circuit contains three matrices (A_1, A_2, A_3). The first and second matrices store the input image, and the third matrix forms the image of the area of intersection of the images of the first two matrices. An example of the formation of the area of intersection of two images is shown in Fig. 7.6. The states of the cells of the third matrix at each time step are shown in Fig. 7.6. Similarly, the cell states at each time-step of the upward shift are shown.

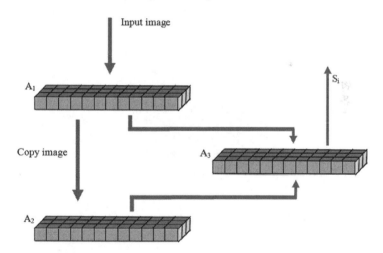

Fig. 7.5: Structure of the device for calculating the intersection of images

Figure 7.6 clearly shows that for a given image the amount of time shifts is less than the number of time shifts to the right.

The structure of the first A_1 matrix consists of homogeneous cells provided with an input and an output (Fig. 7.7).

Each cell is a storage element. At the input of the cell comes the signal of the image point and sets the cell in the corresponding state. A single

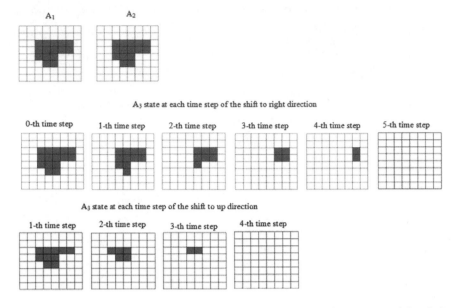

Fig. 7.6: An example of cell states of A_1, A_2, A_3 matrices at each time step of the shift

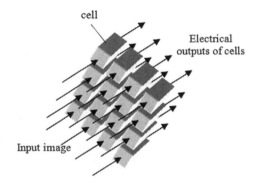

Fig. 7.7: Fragment of the first A_1 matrix

signal sets the cell in a single state and the zero signal sets the cell in the zero state. In fact, the cell of the first A_1 matrix is D-flip-flop.

In the same way, the original image in the second A_2 matrix is recorded. The state of the flip-flop in each cell of the second A_2 matrix is controlled by an additional combinational circuit. In fact, the second A_2 matrix is a two-dimensional reversive shift register that shifts the image in four directions (right, left, up, down).

The structural scheme of the second A_2 matrix is shown in Fig. 7.8.

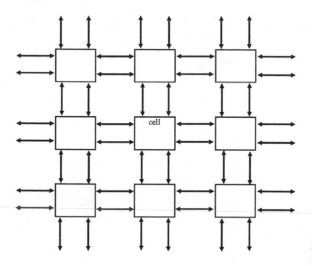

Fig. 7.8: The structure of the second A_2 matrix

The second matrix is represented by an orthogonal lattice. Each cell is associated with four neighbours (two horizontally and two vertically). Such a structure can be constructed on one-dimensional reverse shift registers. Each of the two opposite directions are represented by one layer of one-dimensional reverse shift registers (Fig. 7.9). Information in such registers is recorded in parallel for each category.

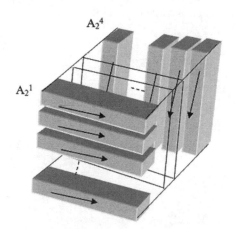

Fig. 7.9: Realisation of the second matrix using one-dimensional reverse shift registers

The most convenient and effective realisation of A_2 is shown in Fig. 7.9. The functional diagram of one cell and four nearest neighbours is shown in Fig. 7.10. Each cell contains a D-flip-flop (T) and a combinational circuit (CC) that controls the state of flip-flop.

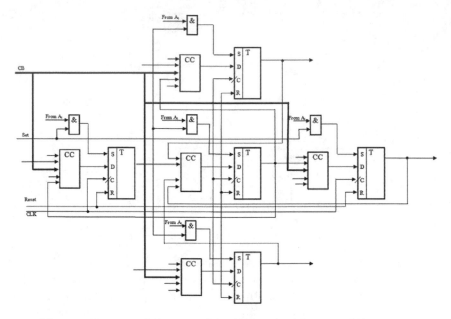

Fig. 7.10: Functional diagram of the fragment of the second A_2 matrix

A functional diagram of one cell with a detailed representation of the combination circuit is shown in Fig. 7.11.

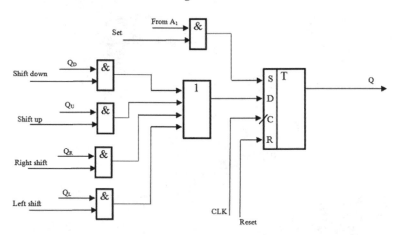

Fig. 7.11: Functional diagram of one cell of the second A_2 matrix

The initial setting of all triggers is performed on the S-inputs. The installation signals come from the outputs of the cells of the first A_1 matrix.

The direction of the shift is carried out by signals on the control bus (CB). The control bus is four-bit. A signal of logic '1' must be present only on one bit of the control bus. Each of the four bits indicates one of the four directions of the information shift. The circuit is not complex and is easily implemented on FPGAs.

The third matrix selects the cells of the first two matrices A_1, A_2, which have a logical state of '1' and have the same coordinates in the plane. The cell of the third matrix is AND gates, whose inputs are outputs of the corresponding cells of the first and second matrices. However, there is a problem of efficient reading of the number of cells of the third matrix that is in the logical '1' state.

There are several options for calculating the number of unit states of the third matrix:

1. Consecutive cell-by-cell scanning of each cell of the third matrix. The number of time-steps corresponds to $N \times M$ ($N \times M$ – matrix size).
2. Sequential analysis of the states of each column (line). The number of time-steps corresponds to N or M.
3. Simultaneous counting of all unit states of the third matrix. It is implemented in one-time steps.

The first two options require a lot of time and the third option can be implemented by non-standard circuitry solutions. An example of such an implementation is shown in Fig. 7.12.

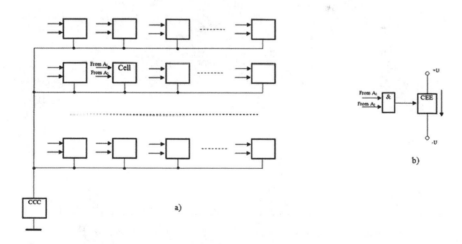

Fig. 7: 12: Example of implementation of the third A_3 matrix in the third variant

The third A_3 matrix of the third embodiment contains $N \times M$ cells. Each cell consumes a fixed value of the current I_1, if both inputs have the logical '1' state. The controlled power element (CEE) consumes the fixed value of the current if its control input has a logic '1' signal from the output of the AND gate (Fig. 7.12b). All currents of the enabled (active) cells of the third matrix flow through the current-code converter (CCC) to the common bus. In the CCC circuit, the consumed current of the matrix flows, and at the output is the current-code converter, a code equivalent to the total current formed. Formation of the code at the output of the current-code converter takes place as a one-time step. However, such a circuit encounters difficulties in implementation on FPGAs.

Consider a circuit that implements the third option without the use of analog-digital transformations. Such a scheme is realised by a large number of logical gates. Let us assume that one column (row) consists of eight cells, that is, the third matrix A_3 has 8×8 dimension. Divide the eight-bit register into two by four digits and implement a scheme for determining the number of units in a four-bit cell. To do this, we constitute a truth table (Table 7.1).

Table 7.1: A truth table for a four-bit counting scheme of the number of ones in the code

X_1	X_2	X_3	X_4	y_1	y_2	y_3	y_4
0	0	0	0	0	0	0	0
0	0	0	1	1	0	0	0
0	0	1	0	1	0	0	0
0	0	1	1	0	1	0	0
0	1	0	0	1	0	0	0
0	1	0	1	0	1	0	0
0	1	1	0	0	1	0	0
0	1	1	1	0	0	1	0
1	0	0	0	1	0	0	0
1	0	0	1	0	1	0	0
1	0	1	0	0	1	0	0
1	0	1	1	0	0	1	0
1	1	0	0	0	1	0	0
1	1	0	1	0	0	1	0
1	1	1	0	0	0	1	0
1	1	1	1	0	0	0	1

A graphic representation of such an element is shown in Fig. 7.13.

The outputs of this circuit are described by the following logical expressions:

$$Y_1 = \overline{X_1} \wedge \overline{X_2} \wedge \overline{X_3} \wedge X_4 \vee \overline{X_1} \wedge \overline{X_2} \wedge X_3 \wedge \overline{X_4} \vee \overline{X_1} \wedge X_2 \wedge \overline{X_3} \wedge \overline{X_4}$$
$$\vee X_1 \wedge \overline{X_2} \wedge \overline{X_3} \wedge \overline{X_4}$$

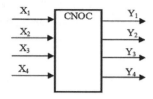

Fig. 7.13: Graphic interface of element counting the number of ones circuit (CNOC) in a four-digit code

$$Y_2 = \overline{X_1} \wedge \overline{X_2} \wedge X_3 \wedge X_4 \vee \overline{X_1} \wedge X_2 \wedge \overline{X_3} \wedge X_4 \vee \overline{X_1} \wedge X_2 \wedge X_3 \wedge \overline{X_4}$$
$$\vee X_1 \wedge \overline{X_2} \wedge \overline{X_3} \wedge X_4 \vee X_1 \wedge \overline{X_2} \wedge X_3 \wedge \overline{X_4} \vee X_1 \wedge X_2 \wedge \overline{X_3} \wedge \overline{X_4}$$
$$Y_3 = \overline{X_1} \wedge X_2 \wedge X_3 \wedge X_4 \vee X_1 \wedge \overline{X_2} \wedge X_3 \wedge X_4 \vee X_1 \wedge X_2 \wedge \overline{X_3} \wedge X_4$$
$$\vee X_1 \wedge X_2 \wedge X_3 \wedge \overline{X_4}$$
$$Y_4 = \vee X_1 \wedge X_2 \wedge X_3 \wedge X_4$$

Such an organisation requires 15 AND gates and 4 OR gates. The number of ones in the code is determined by a logic '1' signal at the corresponding output of the circuit. For example, if a logic '1' signal is present at the second output ($Y_2 = 1$), then there are two logical '1' in the input code. The number of ones in the code is determined by CNOC4 within one-time step.

Now we implement CNOC8, which, for one time step, determines the number of logical '1' in the eight-bit code. For this, an additional combinational circuit is used (Fig. 7.14).

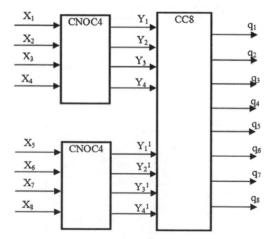

Fig. 7.14: Scheme for calculating the number of ones in the eight-bit code.

The eight-bit combination scheme (CC8) is described by the following logical functions:

$$V = \overline{y_1^1} \wedge \overline{y_2^1} \wedge \overline{y_3^1} \wedge \overline{y_4^1}$$

$$W = \overline{y_1} \wedge \overline{y_2} \wedge \overline{y_3} \wedge \overline{y_4}$$

$$q_1 = y_1 \wedge \overline{y_1^1} \wedge \overline{y_2^1} \vee y_1^1 \wedge \overline{y_1}$$

$$q_2 = y_2 \wedge y_1^1 \vee y_2 \wedge V \vee y_2^1 \wedge W$$

$$q_3 = y_1 \wedge y_2^1 \vee y_2 \wedge y_1^1 \vee y_3 \wedge V \vee y_3^1 \wedge W$$

$$q_4 = y_1 \wedge y_3^1 \vee y_3 \wedge y_1^1 \vee y_2 \wedge y_2^1 \vee y_4 \wedge V \vee y_4^1 \wedge W$$

$$q_5 = y_4 \wedge y_1^1 \vee y_4^1 \wedge y_1 \vee y_2 \wedge y_3^1 \vee y_2^1 \wedge y_3$$

$$q_6 = y_4 \wedge y_2^1 \vee y_4^1 \wedge y_2 \vee y_3 \wedge y_3^1$$

$$q_7 = y_4 \wedge y_3^1 \vee y_4^1 \wedge y_3$$

$$q_8 = y_4 \wedge y_1^1$$

The eight-bit schema will contain 54 logical AND gates and 16 OR gates. This is a small cost within a single crystal of the FPGA. To construct the third matrix, the scheme shown in Fig. 7. 15 is used.

Since the third matrix has a 8×8 dimension, the maximum area of the image can be calculated by 64 cells. An additional combinational circuit (CC64) is used, which has 64 inputs and 64 outputs. Each output indicates the number of cells in a logical '1' state. Each Q_i output is described by

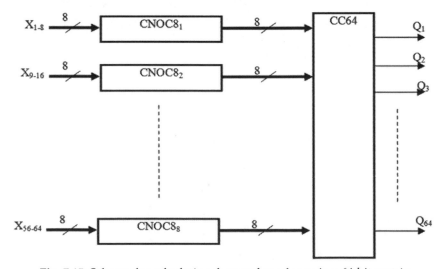

Fig. 7.15: Scheme for calculating the number of ones in a 64-bit matrix

a certain logical expression. These expressions are not given in the text because of their bulkiness. However, this CC64 is synthesised and gives a high performance.

If it is necessary to increase the dimensions of the third A_3 matrix, then it can be created from the already implemented schemes. For large dimensions, this approach complicates the scheme. Therefore, a compromise solution is the parallel calculation of small arrays (for example, array 16×16) and further addition of the received codes.

Thus, a sequence of codes is generated at the output of the UFD, indicating the value of the function of the area of intersection in each time-step. The magnitude of each time-step is determined by the time taken to calculate the number of ones in the third matrix and the time it takes to organise the shift.

CU can be selected as a standard unit. Comparison of codes is widely described in literature.

7.4 Modelling of the Main Hardware Modules for Determining FAI in Modern CAD Systems

There is a need for hardware implementation of an image recognition system based on parallel shift technology. The hardware implementation of such a system can dramatically improve its performance. The proposed method has already been simulated by software and will be discussed in the following sections. Software implementation is time consuming, and thus limits the scope of application.

In the previous paragraphs are given detailed digital diagrams of the main components of the image recognition system based on parallel shift technology. All proposed circuit solutions are presented using a digital element base. Therefore, it is advisable to implement these circuits as embedded systems on FPGAs. For programming FPGA, various software CAD are used (http://www.xilinx.com, http://www.altera.com, http://www.aldec.com).These CAD systems are different in the interface for each firm that manufactures FPGAs.

The most popular are the products of such companies as Xilinx, Altera and others. The authors of the described method work with Xilinx chips and use the software developed by this firm. Xilinx software CAD includes the following (http://www.xilinx.com):

- Alliance Series package, including modules for tracing and interfacing the schematics and text editors, introduced to the project of other firms.
- Foundation Series package, which includes the use of circuit technology, VHDL / Verilog synthesis, functional modelling, crystal tracer and simulation after tracing.

- WebFitter is the FPGA tracer of CPLD of XC9500 series. This product is available only on the Internet and physically located on the server by Xilinx. Only the interface is available to the user.
- WebPack is freely distributed over the Internet for software development HS9500 series FPGAs and CoolRunner and the family of FPGA and Spartan-II of chip Virtex-E XCV300E. Downloading free by the Internet and on PC is installed.
- The Vivado CAD is focused on the intensive use of IP-cores (both provided by Xilinx and other manufacturers, and created by the developer of the project).

All these software products can be downloaded from the Xilinx website (http://www.xilinx.com).

At the same time, any digital circuit can be described using special programming languages (VHDL, Verilog, Abel, etc.). The models described with these languages can be used in almost all software CAD systems. To implement the recognition system, the VHDL was used (Ashenden, 1990; Zimmermann, 1998; Rushton, 1998; Pong, 2006; Ashenden, 2001; Mealy and Tappero, 2012).

Using the VHDL language, a project can be represented by different levels of the hierarchy:

- Behavioral
- Register transfer level
- Functional-logic level using the delay blocks
- Level of logic elements taking into account the switching time
- Level on the basis of individual components of the project, which are stored in a special library

The resulting VHDL code is loaded into the selected CAD, which generates a bitmap file. This file is used to program the selected FPGA.

To prepare the project, several CAD systems were used—Active-HDL, WebPack andVivado. With their help, the main components of the system were obtained in the form of VHDL codes. This approach allows determination of the percentage of use of the main elements of the FPGA. This enables determination of the cost of the project and the timing of the main operations, as well as assessment of the level of energy consumption.

Consider the implementation of the basic elements of the system. The simulation was performed without taking into account the delays at each logical element of the system.

The first element of the simulation is the first A_1 matrix (Fig. 7.5) for stationary storage of the original image. The first A_1 matrix consists of homogeneous memory elements. In accordance with the function performed, the element of the first matrix is represented as an RS flip-flop, which is easily described by VHDL. The model of such an RS-flip-flop is shown in Fig. 7.16.

Fig. 7.16: Graphical interface of RS-flip-flop and time diagrams describing its operation

The first matrix consists of N×M with element RS flip-flops (where N×M is the dimension of the first matrix A_1). Thus, the first matrix consists of N×M inputs and the same number of outputs.

The second A_2 matrix has the same dimension as the first A_1 matrix. Each output of the first matrix is connected to the corresponding input of the second A_2 matrix. In the previous sections, the second matrix A_2 is represented as a two-dimensional shift register. Therefore, for its modelling, a one-dimensional reverse shift register was initially implemented. At first, one bit of such a register was described. Its main element is the controlled D-flip-flop (Fig. 7.17).

This controlled D-flip-flop is the main element of one bit of the second A_2 matrix (Fig. 7.18).

Using the hierarchical principle of creating a VHDL project, a one-dimensional reversive shift register (Fig. 7.19), from which a two-dimensional reverse shift register (Fig. 7.20) were realised. The resulting two-dimensional reverse shift register is the second A_2 matrix.

The resulting time diagrams show the operation of the second A_2 matrix in one direction of the shift. The model has 8×8 dimension and 64 outputs as in the first A_1 matrix.

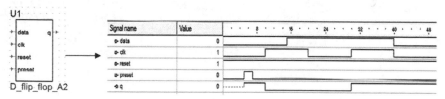

Fig. 7.17: Graphic interface of the controlled D-flip-flop

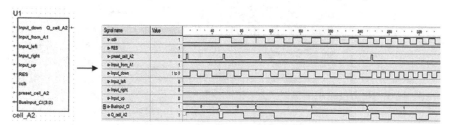

Fig. 7.18: Graphic interface of one digit of the second A_2 matrix and time diagram of its operation

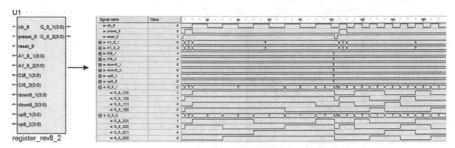

Fig. 7.19: Graphic interface of one-dimensional reverse shift register and time diagrams of its operation

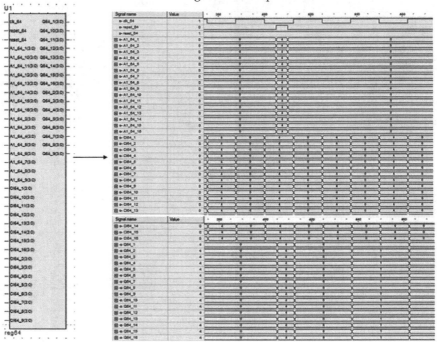

Fig. 7.20: Graphic interface of the second A_2 matrix and time diagrams

The third element of the system is the third matrix A_3, by means of which the function of the area of intersection is determined at each time-step. Each elementary area is determined by a logic '1' signal at the corresponding output of the second A_2 matrix. Each signal of logical '1' comes from the second A_2 matrix. The third matrix A_3 counts all the signals of the logical '1' at its inputs.

The third A_3 matrix was implemented on a hierarchical basis. The first scheme was implemented to count the number of units in the four-digit binary code. From the two four-bit circuits, a circuit was implemented for counting the ones in an eight-bit binary code. These schemes were implemented in accordance with the logical expressions described in the previous paragraph.

In accordance with this approach, a scheme for counting the number of units in a 64-bit binary code was implemented (Fig. 7.21). Its time diagrams are shown in Fig. 7.22.

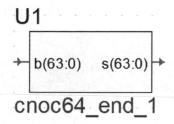

Fig. 7.21: Graphic interface for counting the number of ones in 64-bit binary code

Fig. 7.22: Time diagrams of the counting scheme of the number of ones in a 64-bit binary code

The implementation of such a scheme required a large number of searches of outputs in the 32-bit codes at the input. This approach allows us to construct a scheme for calculating the number of ones in a code of any dimension.

The resulting schemes were implemented in CAD systems, such as WebPack and Vivado, which showed the cost of FPGA elements in the implementation of the system.

7.5 Selection and Storage of Patterns in UMFP

The chief importance lies in the formation of a vector of basic quantitative characteristics that describe the image after the implementation of a parallel shift. It is necessary to generate a form that is easily accessible and understandable and also, effectively stored in the memory of templates. In addition, the generated set of values should not cause problems when comparing. The comparison unit should be simple and perform with a minimal time spent.

In the electronic digital scheme, the received FAI is represented by a code that is divided into fields (Fig. 7.23), which determine the value of the function of the area of intersection at the appropriate time.

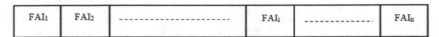

Fig. 7.23: The form of the function of the area of intersection for one of the directions of the shift

The number of functions of the area of intersection depends on the size of the image and the number of discretisation steps of the image field. Typically, a length corresponding to the width (height) of the image field is selected.

When the FAI vector is formed, a vector of the template is searched for, which is to be determined by it. For this, there is the task of efficiently locating and storing the template codes. There are several options for presenting the template codes. The easiest way to represent the codes is shown in Fig. 7.23.

Each field is the time-step number of the shift. The first coordinate is the step number of the shift, and the second coordinate is the code in this field. The search for the template code, which coincides with the code at the input, consists of a sequential search in memory. The implementation of this method consumes a lot of time, which dramatically reduces the performance of the recognition system.

The second option implements the creation of another code that has other meanings than the first version. The code structure is divided into a number of fields (Fig. 7.24). However, each field contains values that have a different value.

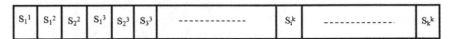

Fig. 7.24: The code structure based on the received function of the area of intersection for one direction of the shift

The code is also divided into code fields. Each field contains a S_i^j value (where i is the number of partitions of the axis of the shift direction and j is the number of the partition field), which determines the curvature of the function of the area of intersection at the chosen increment. For example, the FAI(x) was obtained in the following values (Table 7.2):

Table 7.2: Example of FAI(x) values

FAI(x)	20	15	12	10	8	6	4	3	2	1	0
X	0	1	2	3	4	5	6	7	8	9	10

For this example, we get the following values, which form the template code:

$$S_1^1 = \frac{20}{10} = 2$$

$$S_1^2 = \frac{10-6}{5} = 0.8 \; ; \; S_2^2 = \frac{6-0}{5} = 1.2$$

$$S_1^3 = \frac{20-10}{3} = 3.3 \; ; \; S_2^3 = \frac{10-4}{3} = 2 \; ; \; S_3^3 = \frac{4-1}{3} = 1$$

$$S_1^4 = \frac{20-12}{2} = 4 \; ; \; S_2^4 = \frac{12-8}{2} = 2 \; ; \; S_3^4 = \frac{8-4}{2} = 2 \; ; \; S_4^4 = \frac{4-2}{2} = 1$$

$$S_1^5 = \frac{20-12}{2} = 4 \; ; \; S_2^5 = \frac{12-8}{2} = 2 \; ; \; S_3^5 = \frac{8-4}{2} = 2 \; ; S_4^5 = \frac{4-2}{2} = 1 \; ;$$

$$S_5^5 = \frac{2-0}{2} = 1$$

With such an organisation, the code length may be less than that in the first method. In addition, this code allows determination of the scale changes in the shape of objects, as well as simplifies the process of the template code.

Along with little spending of time, this code requires additional non-complex computing resources. The memory devices with various forms of organisation can also be used.

Fields of vectors of template images and images at the input of the system are formed according to the principle, which is shown in Fig. 7.25.

$V_1=$	$S_1^1(1)$	$S_1^2(1)$	$S_2^2(1)$	$S_1^3(1)$	$S_2^3(1)$	$S_3^3(1)$	....	$S_1^k(1)$	$S_2^k(1)$	....	$S_k^k(1)$
$V_2=$	$S_1^1(2)$	$S_1^2(2)$	$S_2^2(2)$	$S_1^3(2)$	$S_2^3(2)$	$S_3^3(2)$	....	$S_1^k(2)$	$S_2^k(2)$	....	$S_k^k(2)$
.											
.											
.					....				....		
$V_w=$	$S_1^1(w)$	$S_1^2(w)$	$S_2^2(w)$	$S_1^3(w)$	$S_2^3(w)$	$S_3^3(w)$	....	$S_1^k(w)$	$S_2^k(w)$	....	$S_k^k(w)$

Fig. 7.25: Structure of the field comparison in the image recognition system based on the parallel shift technology

The first one compares the S_1^1 parameter and selects those vectors in which coincidences occur within the error margin. Further, in the selected fields of vectors, the fields are compared by the S_1^2 and S_2^2 values and a group with matched mask values is singled out. Thus, there is a sequential sorting of each group of fields. The field group is determined by the superscript in each field. The value of the upper index indicates the number of fields in the group. For example, if the superscript is 3, then the group includes three values S_1^3, S_2^3, S_3^3.

There is a sequential sorting of each group of fields in parallel in all vectors and one vector is selected, which defines the image at the input of the recognition system.

If after several steps of comparing the fields, there is one vector whose field groups still continue to coincide with the fields at the input, then this vector is considered to be closest to the vector at the input of the system. If all the fields of the template vector coincide with the vector at the input, a complete and accurate identification of the image at the input of the system is implemented. If all the fields of the vector do not match, then the decision is made about the proximity of the input vector to the corresponding template vector. For example, the symbol 'E' at the input is close to the vector of the symbol 'F'.

If no template vector is found, then the input vector is assigned an identifier by the teacher and under this identifier is written the template storage unit.

The accuracy of recognition depends on the length of the vector, while the speed depends on the use of parallel methods and algorithms for performing all the operations in the system.

7.6 Optical Image Recognition System Based on Parallel Shift Technology

Modern advances in the use of light for the construction of information processing systems and image recognition can dramatically improve their productivity. There are many advantages of using light for image transmission and processing. These include:

- Possibility of parallel image processing
- Low power consumption
- A small number of influences and parasitic capacitances
- Absence of interaction between light rays
- Absence of radiation in the external environment (high security from external interventions)

The use of such advantages and existing optical and optoelectronic devices allows the authors to realise an optical system for image recognition based on the technology of parallel shift. An optical system for determining the FAI images based on parallel-shift technology is shown in Fig. 7.26.

The system contains an image source (IS), optical image separator (OS), the mirror of initial image (MII), k mirrors of copy image (MCI$_1$, ..., MCI$_k$), k optical transparencies (OT$_1$, ..., OT$_k$), photodetector (FD) and the calculator of area (CA).

The image from IS goes to the optical divider OS and divides into two streams. The first optical stream arrives at MII and on reflecting from it is projected onto FD. This image is the initial image and remains stationary over the entire field of the photodetector. The second optical

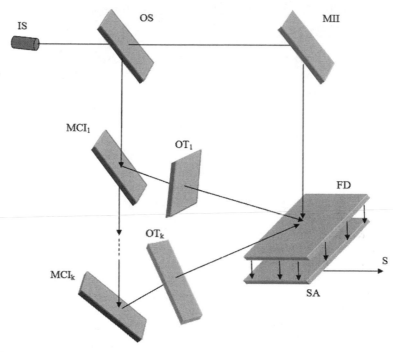

Fig. 7.26: Optical system for calculating the function of the area of intersection of images based on parallel shift technology

flow passes through the translucent mirrors MCI_i and each mirror reflects a portion of the optical flux onto the photodetector FD. In this case, the first MCI_1 is located so that the image is reflected from it and falls on the photodetector as shifted by one discrete image (the unit distance in the selected direction). The second part of the optical stream from MCI_1 is fed to MCI_2. The optical flux from MCI_2 is projected onto a photodetector. In this case, the optical image from MCI_2 falls on the FD as shifted by two single discrete images compared to the original image. MCI_3 realises a shift of the image in the field of the photoreceiver into three discrete intervals, and MCI_k realises a shift of the images in the photodetector field by k discrete intervals.

Each time, only one transparency starts to transmit an optical signal. Between each MCI_i and the photodetector is located one OT_i. Initially, the first optical transparency OT_1 was opened and optical fluxes from the first MCI_1 and from MII come on FD.

At the intersection of both streams, the intensity of the optical signal is greater than the intensity of the optical signal from the MII. The photodetector operates as a threshold element and fixes an area covered by an optical stream with intensity greater than the intensity of the optical signal from MII. The calculator of the area forms the result. This result can be in analog form or in digital form.

The value of the FAI can be written in analog form or in a digital form. Since we considered the digital electronic form earlier, in future we will consider the analog form of storage and representation of the function of the area of intersection.

With the help of Fig. 7.26, the process of image transformation in the function of the area of intersection based on the optical system becomes clear. However, such a system realises the process of obtaining FAI in one direction. The structure of the optical system that forms the function of the area of intersection in all directions is shown in Fig. 7.27.

The second system differs from the first as in the second system fibre optic image distributor (FOID) and four optically controllable transparencies (OCT_0, OCT_{90}, OCT_{180}, OCT_{270}) are added. The second system works in the same way as the first. However, in the second system, there is an additional possibility of rotating the original image by four corners (0°, 90°, 180°, 270°). Image rotation is realised by using FOID. The input original image is fed into the input aperture FOID and distributed over four optical fibres, which are FOID outputs. The output apertures of each fibre are rotated relative to the input aperture FOID, respectively, on 0°, 90°, 180° and 270°. The optical signal is fed to the FOID only from the output of one optical fibre due to the fact that only one optical transparency passes the optical signal.

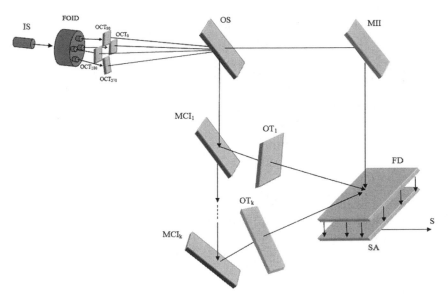

Fig. 7.27: Optical system for formation of FAI images in shear directions

If OST_{90} is transparent, an input image is projected, onto the photodetector which on 90° was rotated. To carry out the shift, the optical transparencies TO, which are located after each MCI, are sequentially switched. This state of OST and OT allows the system to calculate FAI(90°).

Thus, the OST states give the system a shift direction and the OT states realise the image shift in the selected direction. If the number of optical fibres at the output FOID are increased, the number of shifts and the number of received FAI also increase. Optical transparencies can be controlled either optically or electrically, but require rigid alignment of all optical modules of the system.

The task of storing the computed functions of the area of intersection remains. Various variants of optical memory construction are known, and are stored in optical form. Such a memory can be successfully applied to store the values of the function of the area of intersection. However, the authors propose to consider the memory which represents each FAI in the form of a set of magnitudes of optical signal intensities.

Each function of the area of intersection is represented by a line of light emitters (Fig. 7.28).

The recording of the FAI_i in the light-emitting element is carried out by signals on the control inputs (CI). Structure of the light-emitting memory element (LEME) is presented in Fig. 7.29.

The light-emitting memory element comprises a controllable power element (CEE) and a light-emitting element (VD). The controllable power element is included in the same circuit with VD. The installation of the value of the resistance CEE (R_{CEE}) is carried out via the control input

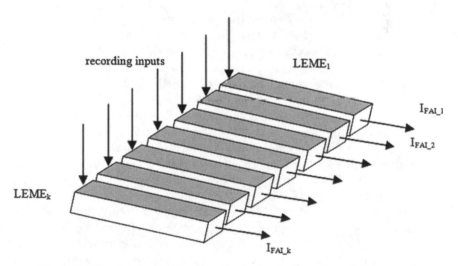

Fig. 7.28: The structure of the line for storing the function of the area of intersection, is computed for one direction.

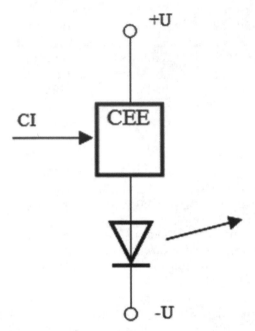

Fig. 7.29: Functional diagram of the light-emitting element of memory

(CI). Resistance R_{CEE} limits the amount of current that flows through VD. Accordingly, VD emits light, which has intensity proportional to the magnitude of the current.

There are many schemes that implement the principle of the controllable power element. The CEE resistance can be controlled by electrical signals, as well as optical signals. In this case, the controllable power element must maintain its resistance even after switching off the supply voltage. After turning on the power, the light-emitting element should emit an optical signal with a fixed intensity.

Thus, the recording of the function of the area of intersection is carried out through the installation of the necessary resistance CEE. This resistance is maintained even when the supply voltage is disconnected. To read the value of the function of the area of intersection, the supply voltage is connected and the light-emitting element emits an optical signal of the corresponding intensity.

Memory of the function of the area of intersection can consist of many layers of memory. One layer of the FAI memory is shown in Fig. 7.30.

Increase in the memory layers allows increase in the amount of memory FAI. Such memory does not require significant power consumption and has a high speed.

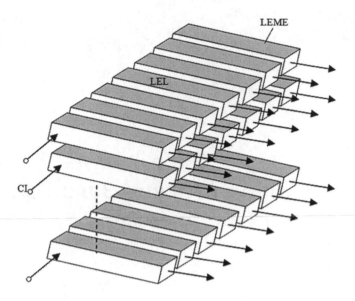

Fig. 7.30: The matrix of memory FAI based on the rulers of light emitters

Methodology of Experimental Studies in Recognition Process Based on Parallel Shift Technology

8.1 Selection and Analysis of Characteristic Features of Images Based on Parallel Shift Technology

The first chapter dealt with the formation and extraction of characteristic features in known methods of image processing and recognition. All these methods can be divided into two classes—the first is the methods determined by the fact that they form the characteristic features by measuring the selected geometric elements of the image. Such elements can be area, perimeter, vertices, angles at the vertices of a polygon, geometric centre, distances between various local image objects, etc.

In fact, the first-class systems measure the quantitative values of the selected geometric elements. From the measured values, a vector of characteristic features is formed. Each characteristic feature is a measured value of the system. Systems of the first class rigidly realise the given functions, for example, almost all known systems for face recognition perform the formation of a vector with characteristic features, which consist of quantities like distance between selected control points. The same approach is used in biometric identification systems where fingerprints are used.

The system of first class consist of software and hardware that measure pre-selected values. However, the vector with characteristic features, consisting only of measured values, does not give a complete description of the images, especially images that are changed in scale or with changed orientation.

The second class of methods is based on the transformation of the original image and the extraction of implicit characteristic features, the quantitative values of which identify only the image. This approach has been given the right to life due to the fact that it gives strict quantitative values that increase the accuracy of image recognition. Such characteristic features include: wavelet transformation coefficients (Daubechies, 1990), potential functions (Aizerman et al., 1970), area intersection functions (Belan and Yuzhakov, 2013; Belan and Yuzhakov, 2013a; Bilan et al., 2014a), pulse sequence characteristics (Belan, 2011; Belan and Belan, 2013), features of wireless sensor networks (Lu et al., 2008; Simek et al., 2011; Heinzelman et al., 2002) and other quantities.

All these values are obtained by using additional transformations that represent the image by an array of quantities. In fact, the geometric characteristics can be represented by a set of quantities that depend on the structure of the converter. This approach can be represented by the following model:

$$I \xrightarrow{\ T\ } F$$

where I is an array of image points that have a certain brightness, colour and geometric shape; T is the function of transformation of input I array; and F is a vector of characteristic features that is obtained due to superimposition transformation of T into I array.

Consider a simple example of image transformation and description. To do this, we choose the Radon transformation, which differs from the methods considered in this book (Belan and Motornyuk, 2013; Bilan et al., 2014). Using the Radon transformation, the image is represented by projections that are received in selected directions. The more projections is the more accurate the image description.

Each projection consists of numbers. Each number is determined by the number of ones on the corresponding line. Each logic '1' encodes the state of a single image point. For example, if the number in the projection is 5, then on the straight perpendicular projection there are five single cells. A set of projection codes describes an input image. Examples of the graphic representation of Radon projections for simple and complex images are shown in Fig. 8.1.

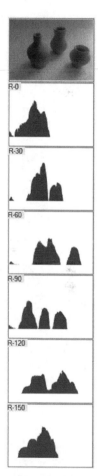

Fig. 8.1: Examples of graphic representation of Radon projections for a brightness level of 37 per cent

For the example presented, the real images were binarised by a brightness level of 50 per cent. After binarisation, Radon projections were obtained.

The Radon transformation is one of a variety of methods for extracting and forming a vector of characteristic features. This method is easily accessible for understanding and is widely used in various applications (Belan and Motornyuk, 2013; Bilan et al., 2014).

Transformation and description of the image based on parallel shift technology can be carried out by two methods:

1. Description of the image using the function of the area of intersection.
2. Description of the image by means of quantitative characteristics of the basic properties of the received FAI.

The first method is based on the construction of an array of numbers that represent the function of the area of intersection. These numbers are obtained by measuring and calculating the function of the area of intersection at each time-step. This is a method of direct measurement of the area, which is a geometric characteristic. Recognition in the first method of description consists of a sequential search and simultaneous comparison of all the FAI numbers with the numbers of the template functions of the area of intersection that are stored in the template storage unit. If there is a case of coincidence in all the FAI numbers, it is decided that the image at the input is the image whose FAI template coincides with the input function of the area of intersection.

The codes of the function of the area of intersection, which are stored as template FAI, have a specific identifier. Also, all recognised and identified images have an identifier belonging to the class.

However, for real images, not all the numbers of the generated FAI at the input can coincide with all the numbers of the template FAI. Herewith, the image can be recognised correctly. For this purpose, the amount of permissible deviations in the quantities is established and the percentage of matched numbers form the FAI code. In addition, the locations of the matched and non-matched numbers in the FAI code are determined. Such values are determined experimentally. The results of the experimental research will be presented in the following sections.

The second method is based on analysis of the form of the received FAI. As described earlier, the function of the area of intersection is a curve in a two-dimensional coordinate system. The FAI curve at the initial coordinate is equal to the maximum value of the FAI_{max}, and at the last coordinate $FAI=0$. The last coordinate corresponds to the value for which $FAI=0$. The last coordinates have different values for different images and for different directions of the shift. The curved line of the function of the area of intersection connects the initial and final coordinates.

Thus, there is at least one of the directions for which the shape of the FAI curve of the image differs from the other image. The shape of the FAI curve is analysed by using the second method. To analyse the shape, the curvature values are used at selected sections of the FAI curve. Curvature is defined by the following formula:

$$R = \frac{\Delta y}{\Delta x}$$

There are many options for analysing the curvature of the FAI line. For a more accurate analysis of the curvature, the bisection method is used (Fig. 8.2).

As can be seen from the example (Fig. 8.2), the curve can be described by the following set of values:

$$R_0 = \frac{\Delta y_0}{\Delta x_0}$$

$$< R_1^1 = \frac{\Delta y_1^1}{\Delta x_1^1}, R_1^2 = \frac{\Delta y_1^2}{\Delta x_1^2} >$$

$$< R_2^1 = \frac{\Delta y_2^1}{\Delta x_2^1}, R_2^2 = \frac{\Delta y_2^2}{\Delta x_2^2}, R_2^3 = \frac{\Delta y_2^3}{\Delta x_2^3}, R_2^4 = \frac{\Delta y_2^4}{\Delta x_2^4} >;$$

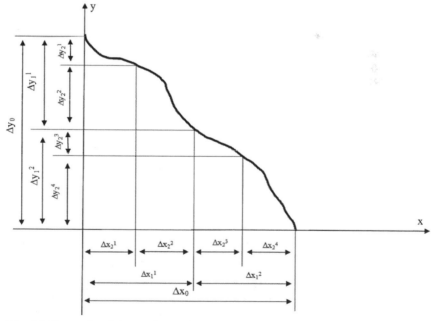

Fig. 8.2: Example of the analysis of the shape of the FAI curve by the method of successive bisection

$$< R_n^1 = \frac{\Delta y_2^1}{\Delta x_2^1}, R_n^2 = \frac{\Delta y_2^2}{\Delta x_2^2}, \ldots\ldots\ldots, R_n^{2^n} = \frac{\Delta y_n^{2^n}}{\Delta x_n^{2^n}} >$$

The accuracy of the analysis of the FAI curve is determined by the number of used curvature values. However, it is important to determine the number of divisions in the values of the X axis. Simply put, it is necessary to determine the maximum number of steps for dividing the segments of the X axis. It is also necessary to determine the maximum number of steps for dividing the segments of the X-axis. It is also a task to determine the length of a segment of the X-axis that satisfies the requirements of an accurate FAI analysis and, as a consequence, image recognition.

The second one is the method that is described in Fig. 8.3.

Such a method (Fig. 8.3) allows dividing the X axis into two segments. However, two segments do not give high accuracy of the FAI analysis. Therefore, all the values obtained are used for each embodiment of the X-axis fission. The method analyses the entire function of the area of intersection. It is necessary to determine the distance of the change in the length of a unit interval ($x_i^j - x_{i-1}^j = \Delta x$).

It is also possible to divide into two segments, four segments, five segments, etc. Other effective variants of analysis are possible. In our

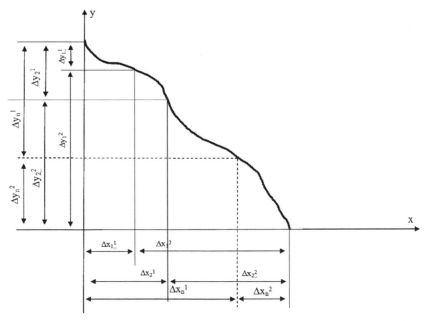

Fig. 8.3: Example of the analysis of the shape of the FAI curve by means of a sequential division into two segments

opinion, an effective analysis can be the method of searching for the largest deviations from the midline or the method of searching for extremes in the FAI graphs.

The task of experimental studies is to determine the sections of the curve that does not significantly affect the recognition result. The presence of such areas and their effective search allows an increase in the performance of the image recognition system based on the technology of parallel shift.

8.2 Analysis of the Accuracy of Image Recognition, Based on the Obtained FAI

The developed methods of image processing and recognition based on the parallel shift technology are simple and not complex in software or hardware implementation. The considered methods of preliminary image processing and noise control confirm the high efficiency in the use of the parallel shift technology. However, there is a need to investigate the use of parallel shift technology directly for image recognition.

In order to carry out the image recognition studies, a computer simulation of the FAI formation method was carried out. Based on the developed program, the function of the area of intersection was formed for direction at the following angles—0°, 45°, 90°, 135°, 180°, 225°, 270° and 315°. To perform the study, images of the BMP format with a size of 500×500 discrete elements (pixels) are selected.

The first area of experimental research was the determination of the deviation of a plane figure at the input from a standard ideal figure. To solve this problem, an ideal reference image was initially set and a lot of images were formed, into which various distortions were artificially introduced. Such distortions were replacement of black pixels with white ones and vice versa. Various variants of substitution were used, which were chosen randomly. A lot of images were formed with distortion of one pixel, with distortion of two pixels, etc.

Such sets were created in order to determine the behavior of deviations from the standard model of the function of the area of intersection. Such an analysis makes it possible to establish the nature of the deviations, which in the future will allow us to realise the functions of a system for an accurate recognition of images of plane figures. For performing of the experimental studies, the images shown in Fig. 8.4 were selected.

These images (Fig. 8.4) were converted into binary ones with an intensity threshold equal to 50 per cent. The obtained values of the function of the area of intersection of these figures were recorded in the memory unit of standard models. The function of the area of intersection of the standard models of figures is presented in Table 8.1.

After the function of the area of intersection of the standard figures are obtained, a set of the functions of the area of intersection for the figures with the introduced changes was formed and the deviations of the FAI values from the standard models are determined. Examples of images that are formed on the basis of standard ones and have distortions are shown in Fig. 8.5.

Fig. 8.4: Images that have been selected as standard models for conducting experimental studies

Table 8.1: The function of the area of intersection of the standard models of figures shown in Fig. 8.4

Figures	Time-step number	FAI for selected shift directions							
		0°	45°	90°	135°	180°	225°	270°	315°
Triangle	1.	45	36	45	45	45	36	45	45
	2.	36	21	36	36	36	21	36	36
	3.	28	10	28	28	28	10	28	28
	4.	21	3	21	21	21	3	21	21
	5.	15	0	15	15	15	0	15	15
	6.	10	0	10	10	10	0	10	10
	7.	6	0	6	6	6	0	6	6
	8.	3	0	3	3	3	0	3	3
	9.	1	0	1	1	1	0	1	1
	10.	0	0	0	0	0	0	0	0
Square	1.	90	81	90	81	90	81	90	81
	2.	80	64	80	64	80	64	80	64
	3.	70	49	70	49	70	49	70	49
	4.	60	36	60	36	60	36	60	36
	5.	50	25	50	25	50	25	50	25
	6.	40	16	40	16	40	16	40	16
	7.	30	9	30	9	30	9	30	9
	8.	20	4	20	4	20	4	20	4
	9.	10	1	10	1	10	1	10	1
	10	0	0	0	0	0	0	0	0

(Contd.)

Table 8.1: (*Contd.*)

	1.	102	95	100	95	102	95	100	95
	2.	91	78	88	78	91	78	88	78
	3.	80	62	76	62	80	62	76	62
	4.	69	47	65	47	69	47	65	47
	5.	58	33	54	33	58	33	54	33
	6.	47	20	44	20	47	20	44	20
Pentagon	7.	36	10	34	10	36	10	34	10
	8.	25	3	25	3	25	3	25	3
	9.	16	0	16	0	16	0	16	0
	10.	9	0	8	0	9	0	8	0
	11.	4	0	0	0	4	0	0	0
	12.	1	0	0	0	1	0	0	0
	13.	0	0	0	0	0	0	0	0

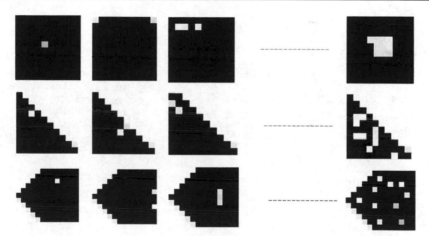

Fig. 8.5: Examples of images that are based on template images with distortions

For images with distortions, the function of the area of intersection was obtained for the same shift directions. Also, the deviations of these FAI distorted images from the template image were calculated and the deviations between the functions of the area of intersection of the distorted images. The results of such calculations are presented in Tables 8.2–8.4.

Table 8.2 presents the results for three variants of distorted images. Images that are distorted in one and seven pixels are represented. The results are also shown graphically in Fig. 8.6.

For the second figures, there are variants of figures in which two and ten pixels were distorted. The results are also shown graphically in Fig. 8.7.

Table 8.2: Results of calculation of deviations for the first template image

Time-step number	0°			45°			90°			135°			180°			225°			270°			315°		
	1st	2nd	3rd	1st	2nd	3rd	1st	2nd	3rd	1st	2nd	3rd	1st	2nd	3rd	1st	2nd	3rd	1st	2nd	3rd	1st	2nd	3rd
1.	2	1	2	2	0	2	2	1	2	2	1	2	2	1	2	2	0	2	2	1	2	2	1	2
2.	2	1	1	2	0	0	2	1	1	2	1	1	2	1	1	2	0	0	2	1	1	2	1	1
3.	2	1	1	2	0	0	2	1	1	2	1	1	2	1	1	2	0	0	2	1	1	2	1	1
4.	2	1	1	2	0	0	2	1	1	2	1	1	2	1	1	2	0	0	2	1	1	2	1	1
5.	1	1	1	0	0	0	1	1	1	0	1	1	1	1	1	0	0	0	1	1	1	0	1	1
6.	0	1	1	0	0	0	0	1	1	0	1	1	0	1	1	0	0	0	0	1	1	0	1	1
7.	0	1	1	0	0	0	0	1	1	0	1	1	0	1	1	0	0	0	0	1	1	0	1	1
8.	0	1	1	0	0	0	0	1	1	0	1	1	0	1	1	0	0	0	0	1	1	0	1	1
9.	0	1	0	0	0	0	0	1	0	0	1	0	0	1	0	0	0	0	0	1	0	0	1	0
10.	0	0	0	0	0	0	0	0	0	0	0	0	0	0	0	0	0	0	0	0	0	0	0	0

Side labels: *Shift direction* (spanning the angle columns); *Variant of deviation* (spanning 1st/2nd/3rd). Left-side row categories: *Number of distortions* (rows 1–4) and *Distortion in one pixel* (rows 5–10).

(Contd.)

Table 8.2: (*Contd.*)

Distortions in seven pixels

	1	2	3	4	5	6	7	8	9	10	11	12	13	14	15	16	17	18	19	20	21	22	23	24
1.	14	5	12	14	9	10	14	5	11	13	8	10	14	5	12	14	9	10	14	5	11	13	8	10
2.	5	5	14	7	9	12	5	5	13	7	8	13	5	5	14	7	9	12	5	5	13	7	8	13
3.	5	5	14	7	9	14	4	5	14	8	8	14	5	5	14	7	9	14	4	5	14	8	8	14
4.	4	5	10	7	7	11	4	5	11	6	7	13	4	5	10	7	7	11	4	5	11	6	7	13
5.	4	4	3	7	5	7	3	3	4	6	6	7	4	4	3	7	5	7	3	3	4	6	6	7
6.	3	2	0	7	5	3	3	2	1	5	6	1	3	2	0	7	5	3	3	2	1	5	6	1
7.	2	2	0	5	5	0	1	2	0	3	6	0	2	2	0	5	5	0	1	2	0	3	6	0
8.	2	2	0	7	5	0	2	2	0	4	6	0	2	2	0	7	5	0	2	2	0	4	6	0
9.	0	1	0	0	3	0	0	1	0	0	4	0	0	1	0	0	3	0	0	1	0	0	4	0
10.	0	0	0	0	0	0	0	0	0	0	0	0	0	0	0	0	0	0	0	0	0	0	0	0

Distortions in ten pixels

	1	2	3	4	5	6	7	8	9	10	11	12	13	14	15	16	17	18	19	20	21	22	23	24
1.	20	8	15	20	13	14	20	8	16	18	13	13	20	8	15	20	13	14	20	8	16	18	13	13
2.	8	8	18	10	12	17	8	8	19	10	12	16	8	8	18	10	12	17	8	8	19	10	12	16
3.	8	7	20	10	11	20	6	7	20	12	11	19	8	7	20	10	11	20	6	7	20	12	11	19
4.	6	7	13	10	10	16	6	6	13	8	10	16	6	7	13	10	10	16	6	6	13	8	10	16
5.	6	5	5	10	8	10	4	4	5	8	8	10	6	5	5	10	8	10	4	4	5	8	8	10
6.	4	3	1	10	8	4	4	3	1	6	8	4	4	3	1	10	8	4	4	3	1	6	8	4
7.	3	3	0	8	7	0	1	3	0	4	7	0	3	3	0	8	7	0	1	3	0	4	7	0
8.	2	2	0	10	7	0	2	2	0	4	7	0	2	2	0	10	7	0	2	2	0	4	7	0
9.	0	1	0	0	4	0	0	1	0	0	5	0	0	1	0	0	4	0	0	1	0	0	5	0
10.	0	0	0	0	0	0	0	0	0	0	0	0	0	0	0	0	0	0	0	0	0	0	0	0

Table 8.3: Results of calculation of deviations for the second template image

| | | 0° | | | 45° | | | 90° | | | 135° | | | 180° | | | 225° | | | 270° | | | 315° | | |
| | | Variant of deviation | | | Variant of deviation | | | Variant of deviation | | | Variant of deviation | | | Variant of deviation | | | Variant of deviation | | | Variant of deviation | | | Variant of deviation | | |
Number of distortions	Time-step number	1st	2nd	3rd	1st	2nd	3rd	1st	2nd	3rd	1st	2nd	3rd	1st	2nd	3rd	1st	2nd	3rd	1st	2nd	3rd	1st	2nd	3rd
	1.	4	1	3	2	1	2	4	3	3	4	2	4	4	1	3	2	1	2	4	3	3	4	2	4
	2.	2	1	3	2	0	2	2	2	3	3	2	4	2	1	3	2	0	2	2	2	3	3	2	4
	3.	1	0	2	1	0	2	2	2	2	3	2	4	1	0	2	1	0	2	2	2	2	3	2	4
	4.	1	0	2	1	0	1	2	2	1	3	2	3	1	0	2	1	0	1	2	2	1	3	2	3
Distortions in two pixels	5.	0	0	1	0	0	0	1	2	0	1	2	1	0	0	1	0	0	0	1	2	0	1	2	1
	6.	0	0	0	0	0	0	1	2	0	1	2	0	0	0	0	0	0	0	1	2	0	1	2	1
	7.	0	0	0	0	0	0	0	2	0	0	2	0	0	0	0	0	0	0	0	2	0	0	2	0
	8.	0	0	0	0	0	0	0	1	0	0	1	0	0	0	0	0	0	0	0	1	0	0	1	0
	9.	0	0	0	0	0	0	0	1	0	0	1	0	0	0	0	0	0	0	0	1	0	0	1	0
	10.	0	0	0	0	0	0	0	0	0	0	0	0	0	0	0	0	0	0	0	0	0	0	0	0

Shift direction

(Contd.)

Table 8.3: *(Contd.)*

Distortions in ten pixels

	19	15	19	16	18	16	15	14	15	18	12	18	19	15	19	16	18	16	15	14	15	18	12	18
1.	12	10	13	13	14	14	8	6	8	12	10	14	12	10	13	13	14	14	8	6	8	12	10	14
2.	10	8	9	10	11	10	3	3	2	11	7	9	10	8	9	10	11	10	3	3	2	11	7	9
3.	7	7	8	6	8	7	1	0	1	8	4	7	7	7	8	6	8	7	1	0	1	8	4	7
4.	3	6	4	3	8	5	0	0	0	6	4	3	3	6	4	3	8	5	0	0	0	6	4	3
5.	2	3	2	2	5	2	0	0	0	4	2	1	2	3	2	2	5	2	0	0	0	4	2	1
6.	0	2	1	1	3	0	0	0	0	1	1	1	0	2	1	1	3	0	0	0	0	1	1	1
7.	0	1	0	1	1	0	0	0	0	1	0	0	0	1	0	1	1	0	0	0	0	1	0	0
8.	0	1	0	0	1	0	0	0	0	0	0	0	0	1	0	0	1	0	0	0	0	0	0	0
9.	0	0	0	0	0	0	0	0	0	0	0	0	0	0	0	0	0	0	0	0	0	0	0	0
10.	0	0	0	0	0	0	0	0	0	0	0	0	0	0	0	0	0	0	0	0	0	0	0	0

Table 8.4: Results of the deviation calculation for the third template image

Number of distortions	Time step number	Shift direction																						
		0°			45°			90°			135°			180°			225°			270°			315°		
		Variant of deviation			*Variant of deviation*			*Variant of deviation*			*Variant of deviation*			*Variant of deviation*			*Variant of deviation*			*Variant of deviation*			*Variant of deviation*		
		1st	2nd	3rd	1st	2nd	3rd	1st	2nd	3rd	1st	2nd	3rd	1st	2nd	3rd	1st	2nd	3rd	1st	2nd	3rd	1st	2nd	3rd
	1.	4	4	2	4	4	2	4	3	4	4	4	2	4	4	2	4	4	2	4	3	4	4	4	2
	2.	4	4	2	3	4	2	3	4	4	3	4	2	4	4	2	3	4	2	3	4	4	3	4	2
	3.	3	4	2	2	4	2	2	4	3	2	4	2	3	4	2	2	4	2	2	4	3	2	4	2
	4.	3	2	2	2	2	1	2	4	3	2	2	2	3	2	2	2	2	1	2	4	3	2	2	2
	5.	2	2	2	2	0	1	2	3	2	1	1	1	2	2	2	2	0	1	2	3	2	1	1	1
	6.	1	2	2	1	0	1	2	1	2	0	0	1	1	2	2	1	0	1	2	1	2	0	0	1
	7.	0	2	2	0	0	0	2	0	1	0	0	1	0	2	2	0	0	0	2	0	1	0	0	1
	8.	0	2	2	0	0	0	1	0	0	0	0	0	0	2	2	0	0	0	2	0	0	0	0	0
	9.	0	1	2	0	0	0	1	0	0	0	0	0	0	1	2	0	0	0	1	0	0	0	0	0
	10.	0	0	2	0	0	0	0	0	0	0	0	0	0	0	1	0	0	0	0	0	0	0	0	0
	11.	0	0	1	0	0	0	0	0	0	0	0	0	0	0	0	0	0	0	0	0	0	0	0	0
	12.	0	0	0	0	0	0	0	0	0	0	0	0	0	0	0	0	0	0	0	0	0	0	0	0
Distortions in two pixels	13.	0	0	0	0	0	0	0	0	0	0	0	0	0	0	0	0	0	0	0	0	0	0	0	0

(Contd.)

Table 8.4: (*Contd.*)

	10	15	20	13	14	20	13	16	20	12	13	20	10	15	13	14	20	13	16	20	12	13	20
1.	9	18	14	13	17	17	10	19	14	12	16	18	9	18	13	17	17	10	19	14	12	16	18
2.	9	19	11	11	20	12	9	20	10	11	19	14	9	19	11	20	12	9	20	10	11	19	14
3.	8	13	8	10	20	12	7	13	7	8	17	10	8	13	10	20	12	7	13	7	8	17	10
4.	6	4	6	9	13	9	6	3	6	8	14	10	6	4	9	13	9	6	3	6	8	14	10
5.	4	1	3	9	7	9	4	0	4	8	10	9	4	1	9	7	9	4	0	4	8	10	9
6.	3	0	2	5	0	5	2	0	0	7	7	6	3	0	5	0	5	2	0	0	7	7	6
7.	1	0	0	5	0	4	1	0	0	5	4	3	1	0	5	0	4	1	0	0	5	4	3
8.	0	0	0	5	0	3	0	0	0	5	1	2	0	0	5	0	3	0	0	0	5	1	2
9.	0	0	0	5	0	0	0	0	0	3	0	2	0	0	5	0	0	0	0	0	3	0	2
10.	0	0	0	0	0	0	0	0	0	1	0	0	0	0	0	0	0	0	0	0	1	0	0
11.	0	0	0	0	0	0	0	0	0	0	0	0	0	0	0	0	0	0	0	0	0	0	0
12.	0	0	0	0	0	0	0	0	0	0	0	0	0	0	0	0	0	0	0	0	0	0	0
13.	0	0	0	0	0	0	0	0	0	0	0	0	0	0	0	0	0	0	0	0	0	0	0

Distortions in ten pixels

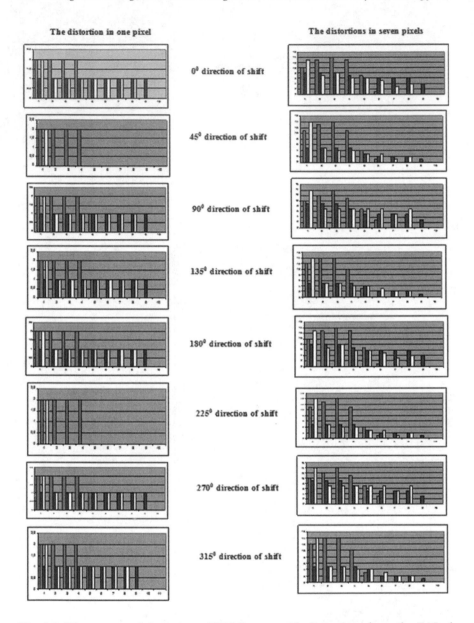

Fig. 8.6: Histograms of deviation of FAI figures with distortions from the FAI of template figures for the first image

For the third figures, variants of figures are presented in which two and ten pixels were distorted. The results are also shown graphically in Fig. 8.8.

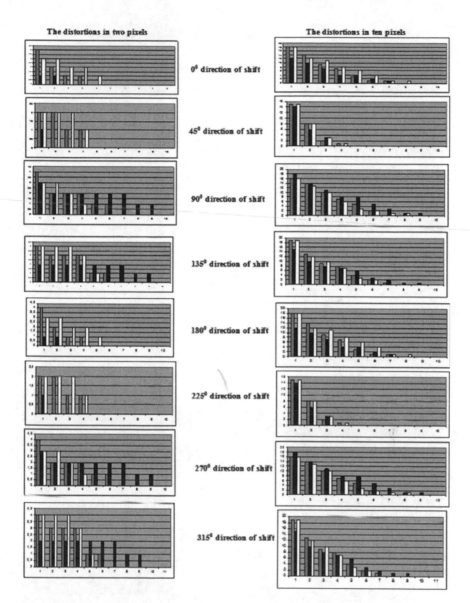

Fig. 8.7: Histograms of the deviation of FAI figures with distortions from the function of the area of intersection of template figures for the second image

Deviations between the functions of the area of intersection of standard figures of different forms were not calculated, since they have different areas and correspondingly different values for the function of the area of intersection.

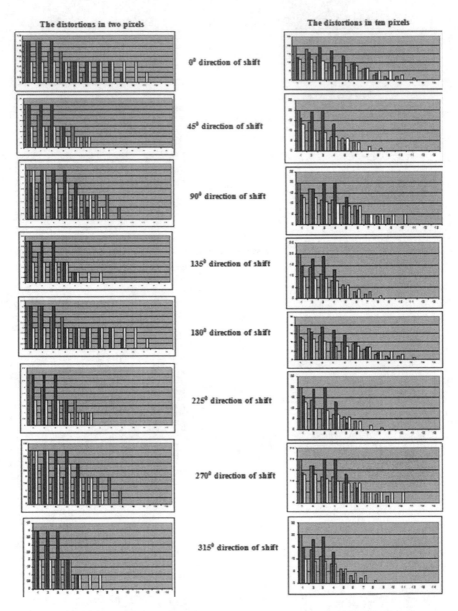

Fig. 8.8: Histograms of deviation of FAI figures with distortions from the function of the area of intersection of template figures for the third image

Based on the results obtained, we see the same behavior of the function of the area of intersection in time. At the first steps of the FAI, the maximum values of deviations are observed. After a certain number of temporary shift steps, these values are reduced to zero value.

The first analysis of the obtained results of the FAI formation in Fig. 8.5, was to determine the amount of deviation of distorted figures from the basic standard figure. As a basic standard model of figure, any figure can be selected. For our experiment, we chose the solid (filled) polygons (square, triangle and pentagon). The convex figures have been selected.

For each template figure, a set of distorted figures is formed. The distortions were replacement of black pixels by white. The functions of the area of intersection for all formed sets of images were also formed. The analysis showed that deviation in any of the possible directions of the shift does not exceed $2n$ (where n is the number of distorted black pixels).

However, this value may not be present in all directions. In the obtained FAI, differences are observed in the first value of the shift. The decrease in differences is determined by the number of pixels P that are located between the edge of the hole and the edge of the figure in the corresponding shift direction (Fig. 8.9).

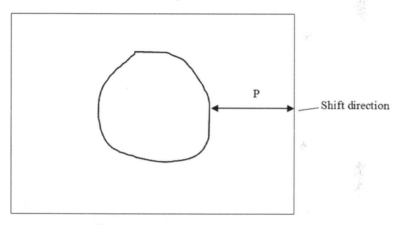

Fig. 8.9: Example of a figure with a hole

After both the holes go beyond the main figure and a copy, the differences is practically not observed. The analysis of the obtained FAI allows determination of the location of the hole or distortion in the figure image. Analysis of all the directions allows determination of the shape of the hole and its dimensions.

Also, analysis can be performed by determining the sequence of quantities that determine the difference between adjacent time shift steps. By the nature of the change in these quantities, we can determine the presence of distortions in the template image. Also, the nature of the change in such quantities allows us to determine the shape of the figure.

An example of changing such quantities for the image of a square is shown in Table 8.5. Table 8.5 shows the squares in which three and five pixels are distorted.

Table 8.5: Results of calculation of deviations for the image of squares at all time steps of shift

			Shift direction																						
			0°			45°			90°			135°			180°			225°			270°			315°		
			Template image	Variant of deviation		Template image	Variant of deviation		Template image	Variant of deviation		Template image	Variant of deviation		Template image	Variant of deviation		Template image	Variant of deviation		Template image	Variant of deviation		Template image	Variant of deviation	
	Time-step number			1st	2nd		1st	2nd		1st	2nd		1st	2nd		1st	2nd		1st	2nd		1st	2nd		1st	2nd
Number of distortions	1.	Distortions in three pixels	10	12	10	19	23	18	10	12	10	19	21	17	10	12	10	19	22	17	10	12	10	19	22	17
	2.		10	11	10	17	17	17	10	11	10	17	18	17	10	11	10	17	17	17	10	11	10	17	18	17
	3.		10	10	10	15	15	15	10	10	10	15	15	15	10	10	10	15	15	15	10	10	10	15	15	15
	4.		10	10	10	13	12	13	10	9	10	13	12	13	10	10	10	13	12	13	10	9	10	13	12	13
	5.		10	7	10	11	7	11	10	8	10	11	8	11	10	7	10	11	7	11	10	8	10	11	8	11
	6.		10	7	10	9	8	9	10	8	10	9	7	9	10	7	10	9	8	9	10	8	10	9	7	9
	7.		10	10	10	7	7	7	10	9	10	7	7	7	10	10	10	7	7	7	10	9	10	7	7	7
	8.		10	10	10	5	5	5	10	10	10	5	5	5	10	10	10	5	5	5	10	10	10	5	5	5
	9.		10	10	9	3	3	2	10	10	9	3	3	3	10	10	9	3	3	2	10	10	9	3	3	3
	10.		10	10	8	1	1	0	10	10	8	1	1	0	10	10	8	1	1	0	10	10	8	1	1	0

(Contd.)

Table 8.5: *(Contd.)*

1.	10	13	11	19	22	17	10	12	10	19	23	17	10	13	11	19	22	17	10	12	10	19	23	17
2.	10	12	10	17	19	17	10	12	10	17	18	17	10	12	10	17	19	17	10	12	10	17	18	17
3.	10	10	10	15	15	15	10	11	10	15	15	15	10	10	10	15	15	15	10	11	10	15	15	15
4.	10	10	9	13	11	13	10	8	10	13	11	13	10	10	9	13	11	13	10	8	10	13	11	13
5.	10	5	9	11	6	11	10	7	10	11	5	10	10	5	9	11	6	11	10	7	10	11	5	10
6.	10	5	10	9	6	8	10	7	10	9	7	9	10	5	10	9	6	8	10	7	10	9	7	9
7.	10	10	10	7	7	7	10	8	10	7	7	7	10	10	10	7	7	7	10	8	10	7	7	7
8.	10	10	10	5	5	5	10	10	10	5	5	5	10	10	10	5	5	5	10	10	10	5	5	5
9.	10	10	8	3	3	2	10	10	8	3	3	2	10	10	8	3	3	2	10	10	8	3	3	2
10.	10	10	8	1	1	0	10	10	7	1	1	0	10	10	8	1	1	0	10	10	7	1	1	0

Distortions in five pixels

Table 8.5 clearly shows the changes in the magnitude of the difference between adjacent shift steps for each FAI. These changes indicate the shape of the figure and its distortions. For greater clarity, the results of such changes for a square with five distorted pixels are shown in Fig. 8.10.

If the image is not solid, but consists of pieces, the differences between the values of the function of the area of intersection at the adjacent steps of the shift may have negative values. An example of such FAI and increments is shown in Fig. 8.11. A complex image is presented which, when intersected with a copy, can give small area values at certain shift steps.

Thus, the analysis of the obtained FAI makes it possible to determine the shape of the figure and the distortions in it. For successful recognition, it is first necessary to form an FAI for reference images and break them down into classes. Based on the available standards, we can determine the shape or pattern, which is filled with a geometric area.

With the help of the first method, it is possible to determine the orientation of the figure, since the FAI figures with different orientations give equal values in most directions. However, in some directions of shear, they differ. An example of such figures and their FAI is shown in Fig. 8.12.

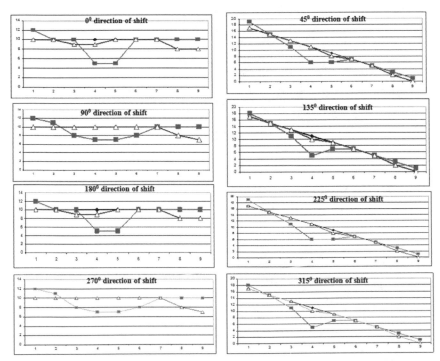

Fig. 8.10: Graphic display of the results of FAI studies according to Table 8.5

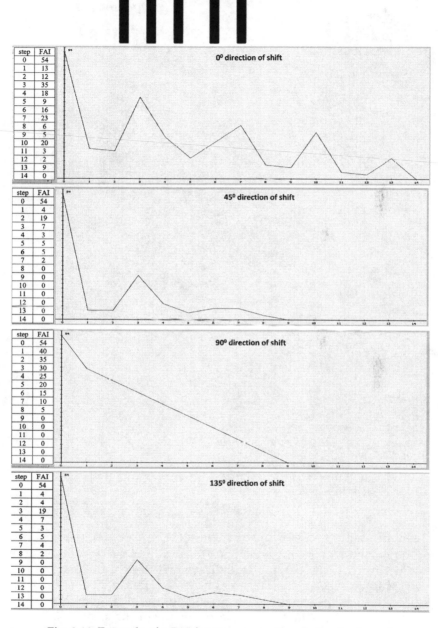

step	FAI
0	54
1	13
2	12
3	35
4	18
5	9
6	16
7	23
8	6
9	5
10	20
11	3
12	2
13	9
14	0

0° direction of shift

step	FAI
0	54
1	4
2	19
3	7
4	3
5	5
6	5
7	2
8	0
9	0
10	0
11	0
12	0
13	0
14	0

45° direction of shift

step	FAI
0	54
1	40
2	35
3	30
4	25
5	20
6	15
7	10
8	5
9	0
10	0
11	0
12	0
13	0
14	0

90° direction of shift

step	FAI
0	54
1	4
2	4
3	19
4	7
5	3
6	5
7	4
8	2
9	0
10	0
11	0
12	0
13	0
14	0

135° direction of shift

Fig. 8.11: Example of a FAI for an image with negative increment
values between shift steps

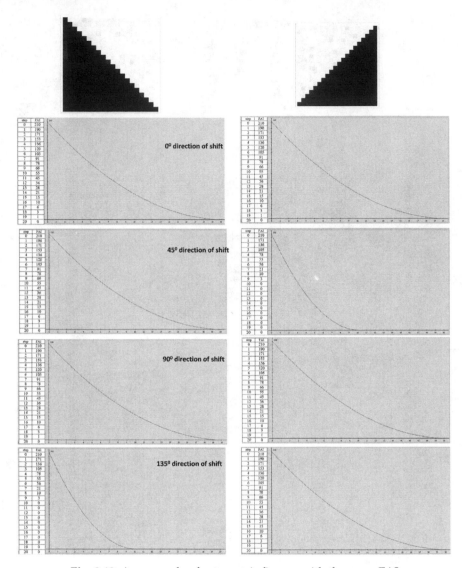

Fig. 8.12: An example of symmetric figures with the same FAI

In the example shown, the functions of the area of intersection differ for the shear directions 45° and 135°. They indicate the orientation of the triangle in the plane. They also indicate that these are equal triangles by magnitude. Changes in the scales of the figure can be determined by the dynamics of the change in the function of the area of intersection. At the same time, the scale factor is determined. For the same images with different scales, there should be a constant coefficient of proportionality of the change in the function of the area of intersection.

8.3 Description of Image with the Help of Quantitative Characteristics of the Basic Properties of FAI

This method differs from the method described in Section 8.2 as in that the FAI curve is analysed. If in the first method the numerical values of the function of the area of intersection are stored and processed, then in the second method, it is the FAI curve.

The graphic method of analysis is described in detail and clearly by using Figs 8.2 and 8.3.

To implement the second method of image description, a program was developed that simulates the technique for analysing the FAI curve. As well as for the previous method, the functions of the area of intersection were formed for shear directions at angles: 0°, 45°, 90°, 135°, 180°, 225°, 270° and 315°. Also for research, images of the BMP format with a size of 500×500 discrete elements (pixels) were selected. The number of division steps was chosen up to seven partitions, that is, the FAI curve can be analysed as a whole, with the help of two sections, three sections, etc. The maximum number of sections of the partition of curve is seven. For larger image sizes, it is necessary to increase the number of sections of the FAI curve. However, if the image has small dimensions, then the possible number of partitions of the FAI curves will decrease.

In the previous section, various deviations from the standard models of images, as well as their orientation were estimated. However, to determine the scale changes using the first method is difficult; most suitable is the second method.

Thus, the first direction of experimental research using the second method is the determination of images of same shapes and different sizes. To solve this problem, a reference image was initially assigned, and its FAI and values of the FAI curvature were determined at all sections of the FAI curve partition. Even images of the same figure were enlarged or reduced in scale.

Thus, sets of figures of the same shape were created. The forms were determined by the number of vertices in the polygon and the angles at these vertices between adjacent sides. For performing experimental studies the images shown in Fig. 8.13 were selected.

For triangles modified on a scale, the following curvature values were obtained for each part of the partition. The results are shown in Figs 8.14–8.16.

Analysis shows that this is a triangle due to the analysis of the 135° direction. The remaining values coincide, which also indicates that this is a right-angled triangle whose cathetus are parallel to the coordinate axes. It is also important to note the dynamics of changes in curvature values of the triangle. For example, for the direction of the shift of 90° and for three

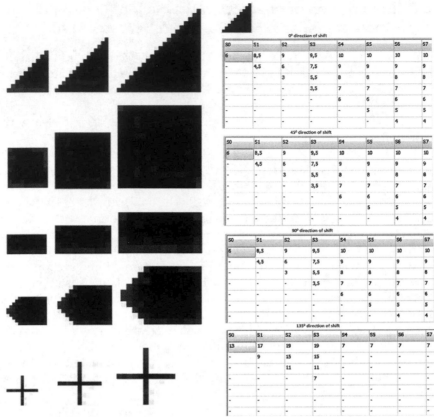

Fig. 8.13: Images selected as standard models for conducting experimental studies by the second method

Fig. 8.14: The results of the FAI curvature analysis for the first triangle

0° direction of shift

S0	S1	S2	S3	S4	S5	S6	S7
6	8,5	9	9,5	10	10	10	10
-	4,5	6	7,5	9	9	9	9
-	-	3	5,5	8	8	8	8
-	-	-	3,5	7	7	7	7
-	-	-	-	6	6	6	6
-	-	-	-	-	5	5	5
-	-	-	-	-	-	4	4

45° direction of shift

S0	S1	S2	S3	S4	S5	S6	S7
6	8,5	9	9,5	10	10	10	10
-	4,5	6	7,5	9	9	9	9
-	-	3	5,5	8	8	8	8
-	-	-	3,5	7	7	7	7
-	-	-	-	6	6	6	6
-	-	-	-	-	5	5	5
-	-	-	-	-	-	4	4

90° direction of shift

S0	S1	S2	S3	S4	S5	S6	S7
6	8,5	9	9,5	10	10	10	10
-	4,5	6	7,5	9	9	9	9
-	-	3	5,5	8	8	8	8
-	-	-	3,5	7	7	7	7
-	-	-	-	6	6	6	6
-	-	-	-	-	5	5	5
-	-	-	-	-	-	4	4

135° direction of shift

S0	S1	S2	S3	S4	S5	S6	S7
13	17	19	19	7	7	7	7
-	9	15	15	-	-	-	-
-	-	11	11	-	-	-	-
-	-	-	7	-	-	-	-
-	-	-	-	-	-	-	-
-	-	-	-	-	-	-	-
-	-	-	-	-	-	-	-

partitions, the neighboring values of the quantities differ by 3, and for four directions, the increment values differ by 2. For all other partitions (90° being the direction of the shift), the difference consists of 1. For other shear directions, the difference between adjacent curvature values is different. These values determine the shape, size and orientation of the triangle.

Based on the results of the analysis of the curvature of the function of the area of intersection of the second triangle, the same dynamics of changes in the curvature values of the function of the area of intersection are seen. However, their values are different. For the second triangle, the increments are greater than for the first; for example, for three partitions (shift direction of 90°), the difference in magnitudes is 4, and for the first triangle, it is 3. This indicates that the second triangle is larger than the first.

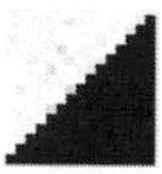

0° direction of shift

S0	S1	S2	S3	S4	S5	S6	S7
8,5	12	13,5	14	14,5	14,5	14,5	15
-	5	9,5	11	12,5	12,5	12,5	14
-	-	5,5	8	10,5	10,5	10,5	13
-	-	-	5	8,5	8,5	8,5	12
-	-	-	-	6,5	6,5	6,5	11
-	-	-	-	-	4,5	4,5	10
-	-	-	-	-	-	2,5	9

45° direction of shift

S0	S1	S2	S3	S4	S5	S6	S7
8,5	12	13,5	14	14,5	14,5	14,5	15
-	5	9,5	11	12,5	12,5	12,5	14
-	-	5,5	8	10,5	10,5	10,5	13
-	-	-	5	8,5	8,5	8,5	12
-	-	-	-	6,5	6,5	6,5	11
-	-	-	-	-	4,5	4,5	10
-	-	-	-	-	-	2,5	9

90° direction of shift

S0	S1	S2	S3	S4	S5	S6	S7
8,5	12	13,5	14	14,5	14,5	14,5	15
-	5	9,5	11	12,5	12,5	12,5	14
-	-	5,5	8	10,5	10,5	10,5	13
-	-	-	5	8,5	8,5	8,5	12
-	-	-	-	6,5	6,5	6,5	11
-	-	-	-	-	4,5	4,5	10
-	-	-	-	-	-	2,5	9

135° direction of shift

S0	S1	S2	S3	S4	S5	S6	S7
17	25	27	29	29	29	29	5
-	13	19	25	25	25	25	-
-	-	11	21	21	21	21	-
-	-	-	17	17	17	17	-
-	-	-	-	13	13	13	-
-	-	-	-	-	9	9	-
-	-	-	-	-	-	5	-

Fig. 8.15: The results of the FAI curvature analysis for the second triangle

0° direction of shift

S0	S1	S2	S3	S4	S5	S6	S7
11	16	17,5	18,5	19	19	19,5	19,5
-	7	11,5	14,5	16	16	17,5	17,5
-	-	5,5	10,5	13	13	15,5	15,5
-	-	-	6,5	10	10	13,5	13,5
-	-	-	-	7	7	11,5	11,5
-	-	-	-	-	4	9,5	9,5
-	-	-	-	-	-	7,5	7,5

45° direction of shift

S0	S1	S2	S3	S4	S5	S6	S7
11	16	17,5	18,5	19	19	19,5	19,5
-	7	11,5	14,5	16	16	17,5	17,5
-	-	5,5	10,5	13	13	15,5	15,5
-	-	-	6,5	10	10	13,5	13,5
-	-	-	-	7	7	11,5	11,5
-	-	-	-	-	4	9,5	9,5
-	-	-	-	-	-	7,5	7,5

90° direction of shift

S0	S1	S2	S3	S4	S5	S6	S7
11	16	17,5	18,5	19	19	19,5	19,5
-	7	11,5	14,5	16	16	17,5	17,5
-	-	5,5	10,5	13	13	15,5	15,5
-	-	-	6,5	10	10	13,5	13,5
-	-	-	-	7	7	11,5	11,5
-	-	-	-	-	4	9,5	9,5
-	-	-	-	-	-	7,5	7,5

135° direction of shift

S0	S1	S2	S3	S4	S5	S6	S7
23	33	35	37	39	39	39	39
-	17	23	29	35	35	35	35
-	-	11	21	31	31	31	31
-	-	-	13	27	27	27	27
-	-	-	-	23	23	23	23
-	-	-	-	-	19	19	19
-	-	-	-	-	-	15	15

Fig. 8.16: The results of the FAI curvature analysis for the third triangle

The analysis of the function of the area of intersection of the third triangle shows that the image of the third triangle is larger than the images of the first two triangles. In the direction of 135°, there are differences and all the partitions are filled. This shift direction indicates the same orientation for images of all triangles.

The results of the analysis of the FAI curve for squares are presented in Figs 8.17–8.19.

An analysis of the function of the area of intersection of the square shows that in the perpendicular shear directions the values of all the curvature coincide. This indicates that the figure is a square with a size of 10×10. The larger squares are represented on Figs 8.18 and 8.19.

From the values obtained, it can be determined that the square has dimensions of 15×15 and more than the previous one.

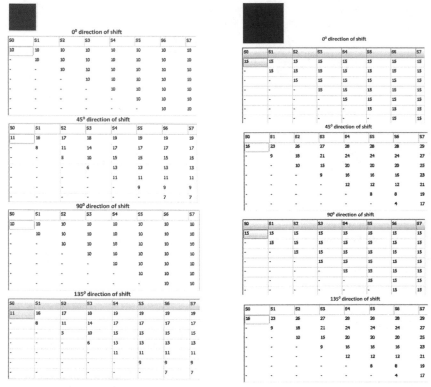

Fig. 8.17: The results of the FAI curvature analysis for the image of the first square (Fig. 8.13)

Fig. 8.18: The results of the FAI curvature analysis for the image of the second square (Fig. 8.13)

The square has dimensions of 20×20 and more than the first two.

The next figures have the rectangles as shown in Fig. 8.13. The results of the studies of the curvature of the function of the area of intersection for these figures are shown in Fig. 8.20.

The obtained values of the curvature for a different number of partitions allow us to judge the size of the figure. For example, we describe quadrangles with dimensions of 10×5, 15×10 and 20×15. From the obtained values, we can determine the exact dimensions of the quadrilateral.

Results were also obtained by studying pentagons. These results describe the figure and determine the differences against the other similar figures. The results of the description of the pentagons are shown in Fig. 8.21.

An analysis of the results obtained show that the differences exist in the last numbers of each column. There is a change in the dynamics in the last steps of the partitions in the horizontal shear directions, since the area on the left side of the figure is smaller than on the right. This allows

0° direction of shift

S0	S1	S2	S3	S4	S5	S6	S7
20	20	20	20	20	20	20	20
-	20	20	20	20	20	20	20
-	-	20	20	20	20	20	20
-	-	-	20	20	20	20	20
-	-	-	-	20	20	20	20
-	-	-	-	-	20	20	20
-	-	-	-	-	-	20	20
-	-	-	-	-	-	-	20

45° direction of shift

S0	S1	S2	S3	S4	S5	S6	S7
21	31	34	36	37	37	38	38
-	13	22	28	31	31	34	34
-	-	10	20	25	25	30	30
-	-	-	12	19	19	26	26
-	-	-	-	13	13	22	22
-	-	-	-	-	7	18	18
-	-	-	-	-	-	14	14

90° direction of shift

S0	S1	S2	S3	S4	S5	S6	S7
20	20	20	20	20	20	20	20
-	20	20	20	20	20	20	20
-	-	20	20	20	20	20	20
-	-	-	20	20	20	20	20
-	-	-	-	20	20	20	20
-	-	-	-	-	20	20	20
-	-	-	-	-	-	20	20

135° direction of shift

S0	S1	S2	S3	S4	S5	S6	S7
21	31	34	36	37	37	38	38
-	13	22	28	31	31	34	34
-	-	10	20	25	25	30	30
-	-	-	12	19	19	26	26
-	-	-	-	13	13	22	22
-	-	-	-	-	7	18	19
-	-	-	-	-	-	14	14

Fig. 8.19: The results of the FAI curvature analysis
for the image of the third square (Fig. 8.13)

Fig. 8.20: The results of the FAI curvature analysis for the image of three
quadrilaterals (Fig. 8.13)

determination of the shape of the figure and the differences against the quadrilateral.

The geometric figures considered are convex. In the following images, more complex figures were selected. Examples of description and analysis of images of crosses are shown in Fig. 8.22.

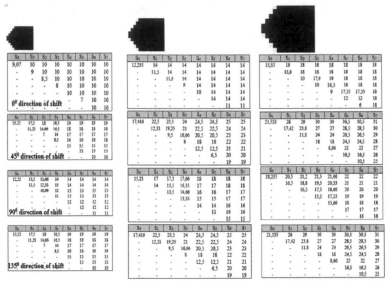

Fig. 8.21: The results of the FAI curvature analysis for the image of three pentagons (Fig. 8.13).

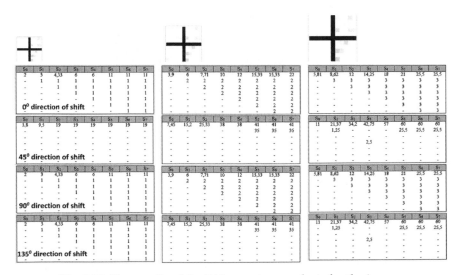

Fig. 8.22: The results of the FAI curvature analysis for the image of three crosses (Fig. 8.13).

For these figures, the number of partitions is sharply reduced in certain shear directions (shear directions 45° and 135°). It is easy to determine the figure from the obtained values, especially, if there is a standard for such a shape.

Let's define the influence of concavities on the result of the image by the second method. For this purpose, images of the following figures were selected (Fig. 8.23). The images of the squares were selected for a clear understanding of the differences as seen against the image of the solid figure. Such an approach makes it possible to determine the influence of concavities on the function of the area of intersection and the curvature of the figure image.

The results of the analysis of the function of the area of intersection of such figures are presented in Figs 8.24–8.26.

In this example, all the values are taken modulo. Therefore, the curvature values are positive with loss of information and reduction in accuracy of image. Therefore, in parallel, it is necessary to analyse the FAI

Fig. 8.23: Images of figures with concavities chosen as standard models for conducting experimental studies on the second method

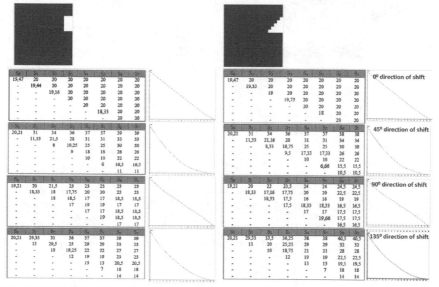

0° direction of shift

S_0	S_1	S_2	S_3	S_4	S_5	S_6	S_7
19,47	20	20	20	20	20	20	20
-	19,44	20	20	20	20	20	20
-	-	19,16	20	20	20	20	20
-	-	-	20	20	20	20	20
-	-	-	-	20	20	20	20
-	-	-	-	-	18,33	20	20
-	-	-	-	-	-	20	20

45° direction of shift

S_0	S_1	S_2	S_3	S_4	S_5	S_6	S_7
20,21	31	34	36	37	37	39	39
-	11,33	21,5	28	31	31	33	33
-	-	8	19,25	25	25	30	30
-	-	-	9	18	18	26	26
-	-	-	-	10	10	22	22
-	-	-	-	-	6	16,5	16,5
-	-	-	-	-	-	11	11

90° direction of shift

S_0	S_1	S_2	S_3	S_4	S_5	S_6	S_7
19,21	20	21,5	23	23	23	23	23
-	18,33	18	17,75	20	20	23	23
-	-	18	18,5	17	17	18,5	18,5
-	-	-	17	19	19	17	17
-	-	-	-	17	17	18,5	18,5
-	-	-	-	-	19	18,5	18,5
-	-	-	-	-	-	17	17

135° direction of shift

S_0	S_1	S_2	S_3	S_4	S_5	S_6	S_7
20,21	29,33	33	36	37	37	39	39
-	13	20,5	25	29	29	33	33
-	-	10	18,25	22	22	27	27
-	-	-	12	19	19	23	23
-	-	-	-	13	13	20,5	20,5
-	-	-	-	-	7	18	18
-	-	-	-	-	-	14	14

0° direction of shift

S_0	S_1	S_2	S_3	S_4	S_5	S_6	S_7
19,47	20	20	20	20	20	20	20
-	19,33	20	20	20	20	20	20
-	-	19	20	20	20	20	20
-	-	-	19,75	20	20	20	20
-	-	-	-	20	20	20	20
-	-	-	-	-	18	20	20
-	-	-	-	-	-	20	20

45° direction of shift

S_0	S_1	S_2	S_3	S_4	S_5	S_6	S_7
20,21	31	34	36	37	37	38	38
-	11,33	21,16	28	31	31	34	34
-	-	8,33	18,75	25	25	30	30
-	-	-	9,5	17,33	17,33	26	26
-	-	-	-	10	10	22	22
-	-	-	-	-	6,66	15,5	15,5
-	-	-	-	-	-	10,5	10,5

90° direction of shift

S_0	S_1	S_2	S_3	S_4	S_5	S_6	S_7
19,21	20	22	23,5	24	24	24,5	24,5
-	18,33	17,16	17,75	20	20	22,5	22,5
-	-	18,33	17,5	16	16	19	19
-	-	-	17,5	18,33	18,33	16,5	16,5
-	-	-	-	17	17	17,5	17,5
-	-	-	-	-	19,66	17,5	17,5
-	-	-	-	-	-	16,5	16,5

135° direction of shift

S_0	S_1	S_2	S_3	S_4	S_5	S_6	S_7
20,21	29,33	33,5	36,25	38	38	40,5	40,5
-	13	20	25,25	29	29	32	32
-	-	10	18,75	21	21	28	28
-	-	-	12	19	19	22,5	22,5
-	-	-	-	13	13	19,5	19,5
-	-	-	-	-	7	18	18
-	-	-	-	-	-	14	14

Fig. 8.24: The results of the FAI curvature analysis for the image of the first and second figures (Fig. 8.23)

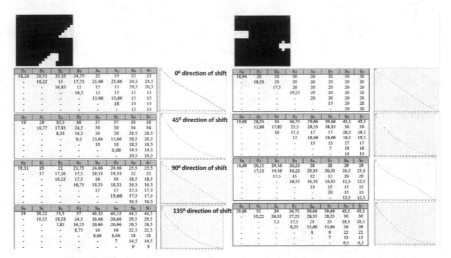

Fig. 8.25: The results of the FAI curvature analysis for the image of the third and fourth figures (Fig. 8.23)

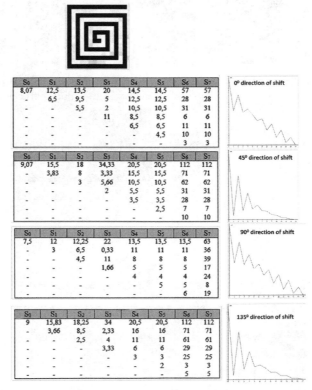

Fig. 8.26: The results of the FAI curvature analysis for the image of the fifth figure (Fig. 8.23)

itself for the presence of negative values of the difference between the neighboring values of the function of the area of intersection. This means that the search negative values are performed. Negative increment values indicate that there are gaps in the image or voids that have a larger area than the elements belonging to the image at each step of the shift. For the fifth figure, the increment values are shown in Table 8.6.

Table 8.6: Results of calculating the deviations for the fifth image (Fig. 8.23) at all time steps of the shift

Time-step number	Shift direction							
	0°	45°	90°	135°	180°	225°	270°	315°
1.	57	112	63	112	57	112	63	112
2.	-28	-71	-36	-71	-28	-71	-36	-71
3.	31	62	39	61	31	62	39	61
4.	-6	-31	-17	-29	-6	-31	-17	-29
5.	11	28	24	25	11	28	24	25
6.	10	-7	-8	-3	10	-7	-8	-3
7.	-3	10	19	5	-3	10	19	5
8.	20	1	-9	7	20	1	-9	7
9.	-11	6	23	0	-11	6	23	0
10.	24	1	-15	6	24	1	-15	6
11.	-13	4	26	0	-13	4	26	0
12.	22	1	-16	4	22	1	-16	4
13.	-9	2	23	0	-9	2	23	0
14.	14	1	-11	2	14	1	-11	2
15.	0	0	14	0	0	0	14	0

More clearly, the dynamics of the change in coordinates can be represented graphically (Fig. 8.27).

Use of the second method allows identification of images that have the same geometric content and are represented in different scales. Also, the second method allows determination of the shape of the figure and the various distortions. Herewith the number of data processed is less than that for the first method, have better visibility and better structuredness.

8.4 Experimental Analysis of FAI of Patterns Image

The parallel shift technology can be effectively used for images that have a complex structure and are used to form various patterns, especially the method for describing and analysing images that are filled with different patterns. Such images have the function of the area of intersection, the shape of which consists of a set of local extremes. A pattern is an image

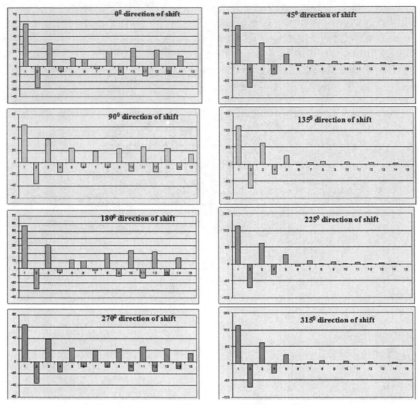

Fig. 8.27: Histograms of the distribution of the values of the change in the FAI values for selected shear directions

consisting of lines, as well as colors and shadows. Patterns also make up templates that are based on repetition and on the alternation of constituent elements. Such patterns form the ornaments.

This form depends on the existence of gaps in the image. The magnitude of the gaps in the image determines the magnitude of the local maxima on the FAI curve. An example of such an FAI is presented for the image of a spiral in Fig. 8.26. Obviously, to estimate the function of the area of intersection by the second method, a large number of partitions and additional computations are required. Therefore, the most appropriate approach for estimating such FAIs is use of a method based on estimating the dynamics of the change in the area values at neighboring time-steps of the shift.

The first indicator of the presence of discontinuities in the image is the presence of a negative number in the code of quantities, which reflect the dynamics of the change in the FAI curve as shown in Table 8.6. The nature of the discontinuities is also determined by the analysis of the FAI obtained in various directions. An example of images with horizontal and vertical discontinuities, as well as their FAIs and the results of calculating the curvature for a different number of partitions are presented in Figs 8.28 and 8.29.

Analysis of the obtained values shows that the first and second methods are applicable for determining the discontinuities in the image. Figure 8.28 shows vertical discontinuity in the direction of shear 90°, since the line of the function of the area of intersection in this direction is almost a straight line. However, in one direction, it is impossible to judge the existence of discontinuity. It is also necessary to analyse the remaining directions. In our example, the functions of the area of intersection in these directions have local maxima, which indicate that there is a gap in the image. On the basis of the function of the area of intersection in the shift directions 0°, 45°, 90° and 135°, it is clear that there is a vertical gap in the image.

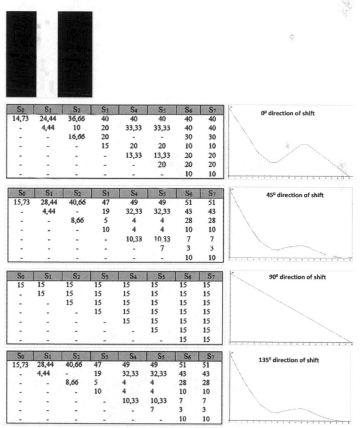

0° direction of shift

S_0	S_1	S_2	S_3	S_4	S_5	S_6	S_7
14,73	24,44	36,66	40	40	40	40	40
-	4,44	10	20	33,33	33,33	40	40
-	-	16,66	20	-	-	30	30
-	-	-	15	20	20	10	10
-	-	-	-	13,33	13,33	20	20
-	-	-	-	-	20	20	20
-	-	-	-	-	-	10	10

45° direction of shift

S_0	S_1	S_2	S_3	S_4	S_5	S_6	S_7
15,73	28,44	40,66	47	49	49	51	51
-	4,44	-	19	32,33	32,33	43	43
-	-	8,66	5	4	4	28	28
-	-	-	10	4	4	10	10
-	-	-	-	10,33	10,33	7	7
-	-	-	-	-	7	3	3
-	-	-	-	-	-	10	10

90° direction of shift

S_0	S_1	S_2	S_3	S_4	S_5	S_6	S_7
15	15	15	15	15	15	15	15
-	15	15	15	15	15	15	15
-	-	15	15	15	15	15	15
-	-	-	15	15	15	15	15
-	-	-	-	15	15	15	15
-	-	-	-	-	15	15	15
-	-	-	-	-	-	15	15

135° direction of shift

S_0	S_1	S_2	S_3	S_4	S_5	S_6	S_7
15,73	28,44	40,66	47	49	49	51	51
-	4,44	-	19	32,33	32,33	43	43
-	-	8,66	5	4	4	28	28
-	-	-	10	4	4	10	10
-	-	-	-	10,33	10,33	7	7
-	-	-	-	-	7	3	3
-	-	-	-	-	-	10	10

Fig. 8.28: An example of an image with a vertical discontinuity and its FAI in shear directions 0°, 45°, 90° and 135°

For the second example (Fig. 8.29), the straight line maps the FAI to the shift direction at 0°. The remaining functions of the area of intersection have a curved shape. The magnitude of the curvature determines the width of the rupture and of the parts of the image that are on either side of the rupture. This is also determined from the values of the curvature on the corresponding sections of the partition.

Both examples were considered for initial understanding of the description of different patterns using parallel shift technology. To describe the technique for recognising different patterns, the images shown in Fig. 8.30 are used.

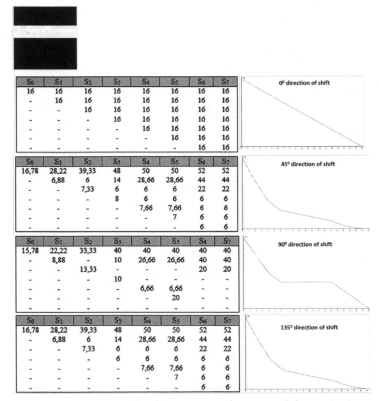

0° direction of shift

S_0	S_1	S_2	S_3	S_4	S_5	S_6	S_7
16	16	16	16	16	16	16	16
-	16	16	16	16	16	16	16
-	-	16	16	16	16	16	16
-	-	-	16	16	16	16	16
-	-	-	-	16	16	16	16
-	-	-	-	-	16	16	16
-	-	-	-	-	-	16	16

45° direction of shift

S_0	S_1	S_2	S_3	S_4	S_5	S_6	S_7
16,78	28,22	39,33	48	50	50	52	52
-	6,88	6	14	28,66	28,66	44	44
-	-	7,33	6	6	6	22	22
-	-	-	8	6	6	6	6
-	-	-	-	7,66	7,66	6	6
-	-	-	-	-	7	6	6
-	-	-	-	-	-	6	6

90° direction of shift

S_0	S_1	S_2	S_3	S_4	S_5	S_6	S_7
15,78	22,22	33,33	40	40	40	40	40
-	8,88	-	10	26,66	26,66	40	40
-	-	13,33	-	-	-	20	20
-	-	-	10	-	-	-	-
-	-	-	-	6,66	6,66	-	-
-	-	-	-	-	20	-	-
-	-	-	-	-	-	-	-

135° direction of shift

S_0	S_1	S_2	S_3	S_4	S_5	S_6	S_7
16,78	28,22	39,33	48	50	50	52	52
-	6,88	6	14	28,66	28,66	44	44
-	-	7,33	6	6	6	22	22
-	-	-	6	6	6	6	6
-	-	-	-	7,66	7,66	6	6
-	-	-	-	-	7	6	6
-	-	-	-	-	-	6	6

Fig. 8.29: An example of an image with a vertical discontinuity and its FAI in shear directions 0°, 45°, 90° and 135°

Fig. 8.30: Examples of images that have a regular pattern and are used as patterns

The function of the area of intersection of submitted images of patterns are shown in Figs 8.31–8.37.

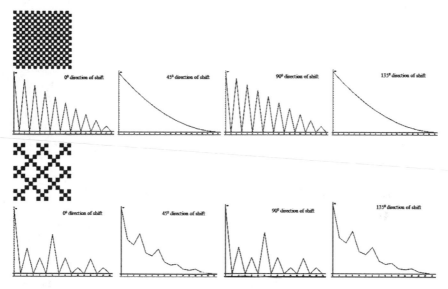

Fig. 8.31: The function of the area of intersection for the first two images is shown in Fig. 8.30 for the shift directions of 0°, 45°, 90° and 135°

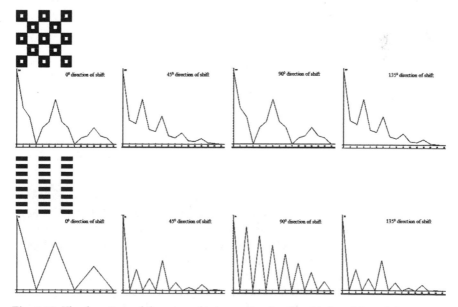

Fig. 8.32: The function of the area of intersection for the third and fourth images is presented in Fig. 8.30 for the shear directions of 0°, 45°, 90° and 135°

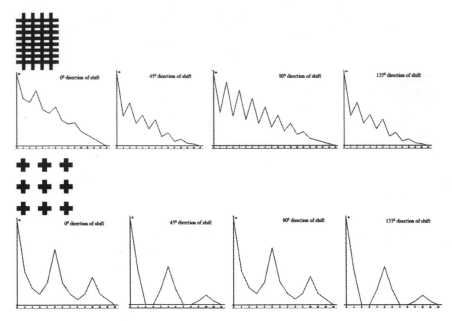

Fig. 8.33: The function of the area of intersection for the fifth and sixth images is presented in Fig. 8.30 for the shear directions of 0°, 45°, 90° and 135°

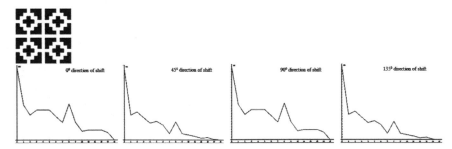

Fig. 8.34: The function of the area of intersection for the seventh image is presented in Fig. 8.30 for the shift directions of 0°, 45°, 90° and 135°

Obviously, all the functions of the area of intersection are represented by curves that have shapes with local extremes. Moreover, these extremes have a certain periodicity until a zero value. For each pattern, the shape of the function of the area of intersection is different from the rest. At the same time, the FAI shapes are the same in the perpendicular directions (0°, 90° and 45°, 135°). However, the function of the area of intersection differs in some areas for the fourth and fifth images. So for the fourth image, we can determine the horizontal coordinates of image discontinuities. At these coordinates, there are no points that belong to the image. The FAI analysis along the directions of the fourth image shows that there are two horizontal ruptures (the number of extrema on the FAI curve for the 0° direction is 2), and the vertical discontinuities are seven (the number of extrema on the FAI curve for the 90° direction is 7). Here we can talk about discontinuities, since the extreme values are quite large. As noted earlier, the presence of local extremes indicates the presence of holes or depressions in the figure image. Discontinuities can only be talked about when there is a zero FAI value between the local extremes (Fig. 8.32).

To describe and recognize such images, it is necessary to calculate the local extremes, the periods of their appearance and the scale changes in the obtained FAI.

8.5 The Sequence of Steps for Using Parallel Shift Technology to Describe and Recognise Images

The proposed image recognition system based on the technology of parallel shift realises an accurate analysis of the function of the area of intersection by quantitative characteristics. An important feature of such a system is the ability to classify images at the input.

Like the classification characteristics of images, topological and geometric characteristics were used. To determine them, an analysis of images of figures with various topological and geometric characteristics was performed. Partially such an analysis is shown in this chapter. Topological characteristics were defined as: the number of vertices of the figure, the number of holes in the filled figure and the number of contour distortions.

The number of vertices is defined as convex vertices (Fig. 8.35a) and concave vertices (Fig. 8.35b). In this case, concave vertices cannot exist without convex vertices. This means that concavity is realised only in convex figures.

In Fig. 8.35(a) is represented the original image, and Fig. 8.35(b) is the same image with two concave vertices. Thus, the number of vertices increased from 6 to 10. Therefore, the first classification characteristic is the number of vertices and all the images of geometric figures organise a

Convex image   Concave vertices

a)   b)

Fig. 8.35: The examples of polygon images

class of polygons. This class is divided into two subclasses—a subclass of images of convex figures and a subclass of polygons with concave vertices.

In Sections 8.2 and 8.3, examples describing images of convex figures and figures that have concavities, are considered. The analysis of FAI forms for such images show how the values of the area vary depending on the number of vertices and their location. This is indicated by the dynamics of the change in the FAI curve. For example, for a pentagon (Fig. 8.21), we see a change in the final part of the FAI curve, which reflects a smaller change in the area at the last steps of the shift, which affects the curvature at the last stages of the partition in the direction of the shift 90°.

Similarly, one can see how the function of the area of intersection changes when there are concavities in the image. In Fig. 8.24 with shift direction of 90°, the FAI curve changes at the last shift steps. Accordingly, the curvature value of the function of the area of intersection at the last step of the partitioning changes. In other directions, the concavity is determined on the right and above.

In comparison to the results presented in Fig. 8.24 and Fig. 8.25, it is possible to determine these concavities by means of parallel shear technology. They represent the same figures with different concavities, as well as the general shapes are preserved, as well as the noticeable changes in the FAI curves in certain areas. Comparative analysis shows that the general forms are preserved, and also that when the time-steps of the shift are observed, the differences between the values of the function of the area of intersection and the curvature are seen. A more detailed analysis of such differences in all directions makes it possible to determine the basic quantitative characteristics of concavities.

Each of the considered classes can be divided into subclasses, which differ in the presence of holes inside the figure. For example, an oval that contains four holes inside can be referred to the class of human faces. Here two holes signify the eyes, one hole is the nose and another hole is the mouth.

The presence of holes was considered earlier (Fig. 8.5). How to determine the presence of a hole and its location is also described above. Detailed analysis of the function of the area of intersection in all directions to determine the exact coordinates of the hole location and its geometric shape is also given. The geometric shape can also be determined by inverting the image. Then the elements that belong to the image become elements of the background, and the elements of the hole become the image. The form of the resulting image is determined by analysing its FAI.

Contour polygon images can also create a separate class of images that are filled with the selected pattern. Patterns are regular expressions that can have gaps. These images and the experimental analysis of their functions of the area of intersection are partially considered in the previous section.

Thus, images can be divided into classes according to the shape of the contour and by the content inside the contour. All these images of geometric shapes can have their names, like planes, houses, windows, faces, people, etc. This applies to individual objects of geometric shapes. If images have complex shapes, then the use of parallel shift technology requires additional preliminary transformations, which are aimed at selection of individual objects in the visual sphere. In future, analysis of the function of the area of intersection of each selected object in the image and the determination of its location is carried out.

According to the conducted research, it is possible to formulate the following recommendations for realisation of the description and recognition of images. The process of description and recognition of images based on the technology of parallel shift can be represented by the following steps:

1. The function of the area of intersection of images at the input of the image recognition system is formed.
2. A set of values of the increments of the FAI curve is formed.
3. The presence of local extremes on the curve of the formed FAI is determined.
4. According to the obtained FAI, the geometric shape of the image of the figure is determined and referred to the corresponding class of polygons.
5. From the obtained values, the curvatures of the FAI curve as per the scale of the recognised triangle is determined.
6. If there are sharp drops or local extremes, then the gaps in the image are determined.
7. According to the analysis of the function of the area of intersection and the curvature of the function of the area of intersection, the presence of holes and their topology is determined.
8. The shape of the FAI curve is determined by the presence of patterns and regular images.
9. If a template image is not found in the base of the standards, then the obtained FAI is assigned an identifier, and it and the curvature values are written into the base of the standards.

The fourth step includes additional operations for recording distortion in the image and distortion of the contour, which can lead to false recognition of the image.

Also, for recognition, the limits of the deviation of the FAI values and the curvature values of the function of the area of intersection are assigned. If the FAI values fall within the permissible limits, then this image is identified by a reference image, where the value of the FAI value lies within these limits. Analogously, the admissible boundaries for the curvatures of the FAI curve are assigned.

If the method of dividing an image into equal parts of a smaller area is used, then the recognition accuracy can be improved. Each section is represented as a separate FAI, which is much smaller and accurately describes the geometric shape in a smaller section. Further, a larger FAI consists of a sequence of the local functions of the area of intersection. Thus, the image can be represented by a number of several types of the functions of the area of intersection—a common FAI for the entire image and an enlarged FAI that is composed of a set of the functions of the area of intersection of individual image regions. The obtained FAI from local FAI allows accurate description and restoration of the image.

The parallel shift technology is effective for processing of large images, regular images and images with discontinuities.

8.6 Advantages of Text Recognition Based on Parallel Shift Technology

The parallel shift technology can be used for almost all types of images. However, there are tasks in the field of image processing and recognition, which are most effectively solved by using the technology of parallel shift. Such tasks of preliminary image processing are discussed in detail in the previous chapters. These are tasks of edge detection, removal of noise, analysis of the trajectory of an object in the visual scene etc.

When recognizing images, effective solutions to the use of parallel shift technology were considered when recognizing images with different orientations and various changes in image sizes. Experiments for solving the described problems are presented as also the examples of simple images. Also the use of parallel shift technology allows better methods to recognise images of patterns and templates. Examples of recognising such images were also examined by using simple and understandable patterns.

However, among all the existing problems of image recognition, there are tasks that can best be solved by using the technology of parallel shift. Such tasks are characterised by the fact that images that are recognised are complex. During the recognition process, complex images are divided into separate elementary elements and each selected image element is described by using the characteristic feature vector.

One of the tasks is of recognising printed and handwritten text. To date, the texts are recognised mainly in the sequence given below.

The process broken into two stages. The first stage is aimed at learning the system. In the process of learning of the image recognition systems, standard models of images of symbols are formed as also their vectors of characteristic features. These codes with identifiers are stored in the memory of the templates codes.

The second stage is the stage of recognition. It consists of the following time-steps:

1. First, each symbol of a text document is selected.
2. The symbol with the help of the characteristic feature vector is described and its code is formed.
3. The resulting code is compared with the codes that are stored in the memory of the patterns.
4. If there is a code in the standard memory that matches the code at the input of the image recognition system, then the input image is recognised. If the pattern code is not founded in the standard memory, the input image of the symbol is not recognised.

This sequence of operations is used in almost all the methods, which have differences mainly in paragraphs 1 and 2 of sequence. There are other methods for selecting the characters in the text (Gonzalez and Woods, 2008; Epshtein et al., 2012; Chen and Yuille, 2004; Lienhart and Wernicke, 2002). These methods do not require implementation of complex algorithms. Also, there are a large number of methods that implement the operation of describing the selected symbol and forming the vector with characteristic features (Gonzalez and Woods, 2008; Chen and Yuille, 2004; Yan and Gao, 2012; González and Bergasa, 2013). Basically, these methods consist of finding characteristic points that distinguish each symbol from the others.

Also, the relationships between the selected characteristic elements of the symbol image are implemented. For example, for the symbol images shown in Fig. 8.36, the characteristic elements are indicated by the circles.

For each characteristic element of the symbol, various algorithms are used to search for them in the image. As can be seen in Fig. 8.36, such characteristic features include elements of intersection of segments, elements that refer to the edges of segments, etc. In this situation, the methods should take into account the thickness of the symbols, which leads to the use of skeletonisation methods for image symbols. This leads to an increase in the time spent on the formation of the vector of characteristic features of each symbol.

At the same time, all these methods are based on a preliminary visual analysis of the topology of each image of character. On the basis of this analysis, the characteristic features of each symbol are chosen. To this end,

Fig. 8.36: Images of characters with selected characteristic elements for each symbol

the developer analyses the images of all symbols and forms a standard symbol base that is used for recognition. Also, the developer implements methods for selecting symbols and extracting their characteristic features.

However, this approach severely limits the image recognition system. This is indicated by several factors. First, if the system is learned by a certain set of symbols, then there is appearance of the image of another symbol (for example, the symbol belongs to another alphabet) and the system can not recognise it. Thus, it is necessary to further study the system.

If the system implements generalised algorithms for selecting characteristic elements, then there are images of symbols that may not have such characteristic elements. For example, the image of the symbol 'O' does not have such characteristic elements that are not displayed in the images of the symbols in Fig. 8.36. In this position, we need to develop additional algorithms.

Symbol recognition of a text takes a lot of time, especially if a text document has a large volume.

Using the methods to recognise a text document by words requires a complete revision of all the existing methods and the development of new methods. One of the methods that realises the recognition of text documents by words is based on the parallel shift technology.

The use of parallel shift technology for recognition of text documents is done as follows:

For each image of a symbol or word that comes to the input of the image recognition system, the standard FAIs are formed in the form of codes that are written into the memory of the standards. In this case, the image recognition system is not trained in advance. Template FAIs are formed during the formation of an image of an unknown symbol or word on the input of the recognition system.

In addition, for the formation of a template FAI, no preliminary visual analysis of the image of each symbol or word is required. That is, there is no need for a preliminary search and selection of characteristic features, which make such a method universal. In fact, use of the technology of parallel shift allows recognition of images of symbols and their combinations of any shape without preliminary analysis. In this case, the number of characters that form a word or number can be any.

The base of pattern codes can be implemented in several ways. The first method forms the base of the pattern codes of large volumes. Here each symbol and each combination of symbols is defined by its own pattern code.

The second method consists of generating the codes of pattern FAI for single-symbol images, and their combinations are represented by the identifier code. If the symbols and their combinations are defined by the

FAI code, these symbols indicate the corresponding identifier in the base of the standards.

Consider the results of an analysis of experimental studies on examples of images of several standard symbols and their combinations.

To carry out experimental studies in problems of recognition of text documents based on the technology of parallel shift, the following conditions and initial settings were chosen. The circumstances taken into consideration was the presence of intersection of images of different shapes when implementing parallel shift. It also took into account a text document with block letters where there are gaps (blanks) between adjacent symbols. Each character in the text is defined by its own sign place. This means that during the formation of a group of symbols, there will be local extreme on the FAI curve. The number of such extrema (local minima or local maxima) indicates the number of characters that constitute the symbol group.

Much attention has been paid to the identification of the basic properties of the shape of the FAI curve. The task was to determine the influence of each symbol on the formation of the FAI of the whole group of symbols.

To solve this problem, images of individual symbols and their combinations with various variants of displacement and permutations in the group were chosen. The images of symbols and their combinations are shown in Fig. 8.37.

Combinations of symbols with large spaces (Fig. 8.37b) were selected in order to determine the function of the area of intersection, which determines the intersection of the unequal shapes. For this purpose, spaces between images of two adjacent symbols were chosen which had a width greater than the width of the symbols. Such images make it possible to determine the intersection of images of two figures of different shapes for two variants. The first variant defines the intersection of the symbols '5' and '7' when the symbol '5' moving to the right covers the symbol '7'. The second option defines the intersection of images of two characters, when the image of the symbol '7' moves to the right and covers the image of the symbol '5'.

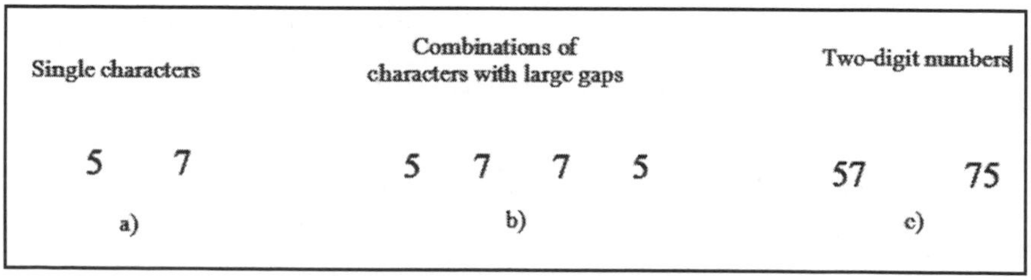

Single characters		Combinations of characters with large gaps				Two-digit numbers	
5	7	5	7	7	5	57	75
	a)			b)			c)

Fig. 8.37: Images of symbols and their combinations, which are established in the experiment are shown

The graphs of the FAI (0°) for the images of the symbols '5' and '7' are shown in Fig. 8.38.

The FAI (0°) curves for each symbol differs both in quantitative characteristics and shape of the FAI curve. These FAI (0°) curves are obtained for a separate image of each symbol, that is, the FAI (0°) is obtained by intersecting the symbol image with its copy. With the help of such separate FAI symbols, one character is recognised.

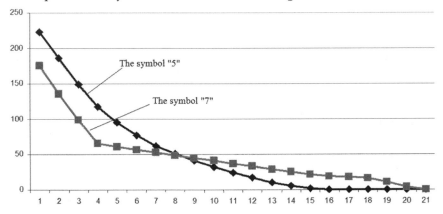

Fig. 8.38: The graphical representation of the FAI (0°) for the images of the symbols '5' and '7' displayed in Fig. 8.37a

If a group of characters (a large number of words in text) is present at the input of the recognition system, then there is a situation when the FAI of two images of different symbols are determined. The first function of the area of intersection is determined for the first variant of the shift (the image of the symbol '5' is shifted towards the image of the symbol '7' from left to right) and the second FAI is determined for the second variant (the image of the symbol '7' is shifted towards the image of the symbol '5' from left to right). The FAI curves for both the variants for the images of the symbols '5' and '7' are shown in Fig. 8.39.

From the obtained FAI (0°) graphs (Fig. 8.39) it is clear that they have an inverse mirror shape. For these forms, we can determine which character was first and which second character (57 or 75). The FAI curves (0°) shown in Fig. 8.39 allow definition of numbers or words that consist of two digits or two characters.

The FAI (0°) for the two symbols constituting the image of number 57 and number 75 differ from the FAI (0°) curves shown in Figs 8.38 and 8.39. However, the endpoints of the FAI (0°) curve for the image of number 57 coincide in shape and in quantitative characteristics with the shape of the FAI (0°) curve for the first variant shown in Fig. 8.39. The end of the FAI (0°) curve for the image of number 75 coincides in form and shape of the

FAI (0°) curve for the second variant as shown in Fig. 8.39. The FAI (0°) curves for the images of numbers '57' and '75' are shown in Fig. 8.40.

The selected endings of FAI (0°) curves clearly show the images of numbers that are present at the input of the recognition system. Table 8.7 presents the quantitative characteristics of FAI (0°) for the FAI obtained in images of the symbols '5' and '7' (5 → 7), as well as '7' and '5' (7 → 5). Also quantitative characteristics for FAI (0°) images of numbers '57' and '75' are presented.

The analysis of FAI (0°) curves makes it possible to recognise the image of the number at the input of the recognition system. However, analysis of the FAI (90°) curve allows a drastic reduction in the time taken to recognise the images of the numbers. Table 8.8 presents the FAI (90°) for the images of symbols '5' and '7', as well as images of numbers '57' and '75'.

Analysis of Table 8.8 clearly shows that the FAI (90°) for images of numbers '57' and '75' are equal. The FAI (90°) for images of numbers '57' and '75' are equal to the sum of the FAI (90°) images of symbols '5' and

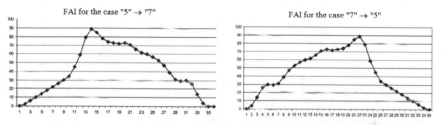

Fig. 8.39: Graphic representation of the FAI (0°) for two variants of intersections of the symbols '5' and '7'

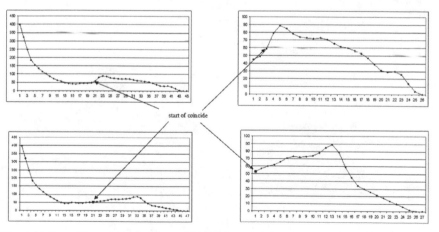

Fig. 8.40: Graphic representation of FAI (0°) curves for images of numbers '57' and '75' and their endings

Table 8.7: The quantitative values of the FAI (0°) for images of the intersections of symbols '5' and '7', '7' and '5', as well as for images of the numbers '57' and '75'

Step number of the image shift	FAI (0°)			
	Image intersection of characters '5' and '7' (5→7)	Image intersection of characters '7' and '5' (7→5)	Image of number '57'	Image of number '75'
0	0	0	398	398
1	2	4	322	322
2	6	14	248	248
3	10	26	184	184
4	14	30	156	156
5	18	29	134	134
6	22	31	115	115
7	26	39	100	100
8	30	47	86	86
9	34	53	73	73
10	45	57	63	61
11	59	60	56	50
12	79	62	49	43
13	89	66	44	44
14	85	71	42	50
15	78	73	41	49
16	74	72	44	47
17	73	73	46	47
18	72	74	45	50
19	73	78	49	51
20	71	85	59	53
21	66	89	79	57
22	62	79	89	60
23	60	59	85	62
24	57	45	78	66
25	53	34	74	71
26	47	30	73	73
27	39	26	72	72
28	31	22	73	73

(Contd.)

Table 8.7: (*Contd.*)

29	29	18	71	74
30	30	14	66	78
31	26	10	62	85
32	14	6	60	89
33	4	2	57	79
34	0	0	53	59
35	0	0	47	45
36	0	0	39	34
37	0	0	31	30
38	0	0	29	26
39	0	0	30	22
40	0	0	26	18
41	0	0	14	14
42	0	0	4	10
43	0	0	0	6
44	0	0	0	2
45	0	0	0	0

Shaded areas of the same color indicate the coincidence of the FAI (0°) regions for different images. These sections clearly show the presence and location of images of symbols in the image of the number.

'7'. Also, the analysis of the data from Table 8.8 and other experimental results shows that for FAI (90°) groups of symbols the following relation holds:

$$FAI_k(90°) = \sum_{i=1}^{k} FAI_k^i(90°), (i = \overline{1,k}),$$

where $FAI_k(90°)$ is function of intersection of the areas of the k-th group of symbols (word, number, etc.) for the direction of shift 90°; and $FAI_k^i(90°)$ is the function of the area of intersection of the i-th symbol, which makes up the k-th group of symbols. For example, for the image number 1273421 following formula is used:

$$FAI_k(90°) = 2 \cdot FAI_7^1(90°) + 2 \cdot FAI_7^2(90°) + FAI_7^3(90°) + FAI_7^4(90°) + FAI_7^5(90°)$$

The superscript indicates the number of the character in the sequence, for example, the first character of the sequence is the image of '1' digit. Since there are two such symbols in the sequence, it is multiplied by two. The same is done with the second character (the image of the '2' symbol) of the sequence. The third character in the sequence (the image of the '7'

Table 8.8: Numerical values of the FAI (90°) for images of symbols '5' and '7', and images of the numbers '57' and '75'

Step number of the image shift	FAI (90°)			
	Image of character '5'	Image of character '7'	Image of number '57'	Image of number '75'
0	223	175	398	398
1	173	143	316	316
2	126	112	238	238
3	83	83	166	166
4	51	56	107	107
5	38	43	81	81
6	31	34	65	65
7	31	25	56	56
8	33	16	49	49
9	36	12	48	48
10	41	12	53	53
11	43	12	55	55
12	41	12	53	53
13	40	12	52	52
14	36	12	48	48
15	33	12	45	45
16	31	12	43	43
17	28	12	40	40
18	28	12	40	40
19	32	12	44	44
20	37	12	49	49
21	40	12	52	52
22	40	12	52	52
23	34	11	45	45
24	27	11	38	38
25	22	11	33	33
26	18	10	28	28
27	18	11	29	29
28	19	11	30	30
29	22	10	32	32
30	23	11	34	34
31	20	11	31	31
32	15	8	23	23
33	7	6	13	13
34	2	3	5	5
35	0	0	0	0

symbol) occurs only once and so are the fourth and fifth characters of the sequence (images of the '3' and '4' symbols).

In Table 8.9, the FAI (90°) for the image 'and' and images of symbols 'a', 'n' and 'd' are presented while in Table 8.10 the FAI (90°) is shown for displaying the group of symbols 'the' images of the individual symbols 't', 'h' and 'e'. A graphic representation of such FAI (90°) is shown in Fig. 8.41.

Table 8.9: Numerical values of the FAI (90°) for images of 'a', 'n', 'd' symbols and the 'and' symbol group

Step number of the image shift	FAI (90°)			
	Image of 'a' character	Image of 'n' character	Image of 'd' character	Image of 'and' symbol group
0	96	117	135	348
1	73	98	113	284
2	62	85	100	247
3	55	77	90	222
4	50	69	81	200
5	45	61	73	179
6	40	53	66	159
7	35	45	62	142
8	30	37	56	123
9	22	29	49	100
10	15	23	43	81
11	10	18	37	65
12	2	10	31	43
13	0	0	23	23
14	0	0	19	19
15	0	0	15	15
16	0	0	12	12
17	0	0	8	8
18	0	0	3	3
19	0	0	0	0

Table 8.10: Numerical values of the FAI (90°) for images of
't', 'h', 'e' symbols and 'the' symbol group

Step number of the image shift	FAI (90°)			
	Image of 't' character	Image of 'h' character	Image of 'e' character	Image of 'the' symbol group
0	70	140	75	285
1	59	121	53	233
2	52	109	39	200
3	47	101	30	178
4	43	93	27	163
5	39	85	25	149
6	35	77	22	134
7	31	69	18	118
8	27	61	12	100
9	23	53	10	86
10	21	45	9	75
11	18	40	7	65
12	13	32	4	49
13	8	24	0	32
14	5	20	0	25
15	3	16	0	19
16	1	13	0	14
17	0	10	0	10
18	0	6	0	6
19	0	0	0	0

The results indicate that it is possible to determine which characters belong to the group. However, they do not allow determination of the sequence of location. The sequence of the arrangement of symbols in the group makes it possible to determine the function of the area of intersection for directions 0°, 45° and 135°.

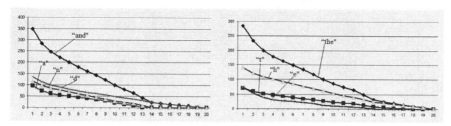

Fig. 8.41: Graphic display of the FAI (90°) curves for the symbol images described in Tables 8.9 and 8.10

Let us consider the FAI (45°) and the FAI (135°), which are obtained for symbols '5', '7' and the numbers '57' and '75'.

The FAI (45°) values for the images of the symbols '5' and '7', as well as for images of the numbers '57' and '75' are shown in Table 8.11.

In Table 8.11, in the sixth column, a notation is introduced that describes the following formula and corresponds to the sum of two FAIs (45°) for images of the symbols '5' and '7'.

$$\Sigma_{5,7} = FAI^5(45°) + FAI^7(45°)$$

The shaded areas of Table 8.11 indicate matches. It also shows that up to the eighth step of the FAI (45°) the image of the number '57' is equal to the sum of the FAI (45°) images of the symbols '5' and '7'. These sums continue until the moment when the FAI (45°) for the symbol of '7' becomes zero. This suggests that in the image of the number '57' the second symbol is '7'. The FAI (45°) for the image of the number '57' has values greater than zero for 35 time-steps of the shift.

The FAI (45°) for the image of the number '75' also has non-zero values of over 35 time-steps of the shift. The values of the FAI (45°) for the image of the number '75' are equal to the sum of the FAI (45°) for the images of symbols '5' and '7', during the 17 time-steps of the shift. This indicates that the image of the symbol '7' starting from the eighth time-step of the shift enters the gap between the symbols '7' and '5'. From the eighth to the seventeenth time-shifting steps, the image of the symbol '7' is placed in the space between the symbols '5' and '7'. Starting from the eighteenth time-step of the shift, the image of the symbol '7' begins to intersect with the image of the symbol '5' (from left to right). This indicates that the image of the symbol '7' is located to the left of the symbol '5'. The FAI (45°) curves for the images of the numbers '5', '7', '57' and '75' are shown in Fig. 8.42.

The FAI (45°) analysis presented in Table 8.11 showed that recognition is affected by the width of the space between the symbols. The FAI (45°) curves of the numbers '57' and '75', until a certain step, merge into one and then bifurcate into separate curves. Prior to this, the FAI (45°) of the

Table 8.11: The numerical values of the FAI (45°) for images of symbols '5', '7' and numbers '57' and '75'

Step number of the image shift	FAI (45°)				$\Sigma_{5,7}$	$\Sigma_{5,7} -$ FAI^{57}(45°)	$\Sigma_{5,7} -$ FAI^{75}(45°)
	Image of '5' character	Image of '7' character	Image of number '57'	Image of number '75'			
0	223	175	398	398	398	0	0
1	162	133	295	295	295	0	0
2	103	93	196	196	196	0	0
3	46	54	100	100	100	0	0
4	19	16	35	35	35	0	0
5	17	4	21	21	21	0	0
6	19	2	21	21	21	0	0
7	26	1	27	27	27	0	0
8	31	0	31	31	31	0	0
9	37	0	38	37	37	-1	0
10	35	0	37	35	35	-2	0
11	28	0	33	28	28	-5	0
12	23	0	31	23	23	-8	0
13	19	0	31	19	19	-12	0
14	19	0	35	19	19	-16	0
15	14	0	33	14	14	-19	0
16	7	0	27	7	7	-20	0

(Contd.)

Table 8.11: *(Contd.)*

	1	0	20	1	1	-19	0
17	0	0	20	1	1	-19	0
18	0	0	17	2	0	-17	-2
19	0	0	16	8	0	-16	-8
20	0	0	15	16	0	-15	-16
21	0	0	14	24	0	-14	-24
22	0	0	13	29	0	-13	-29
23	0	0	12	26	0	-12	-26
24	0	0	13	21	0	-13	-21
25	0	0	15	15	0	-15	-15
26	0	0	18	10	0	-18	-10
27	0	0	21	11	0	-21	-11
28	0	0	24	11	0	-24	-11
29	0	0	33	10	0	-33	-10
30	0	0	40	11	0	-40	-11
31	0	0	42	11	0	-42	-11
32	0	0	33	8	0	-33	-8
33	0	0	19	6	0	-19	-6
34	0	0	7	3	0	-7	-3
35	0	0	0	0	0	0	0

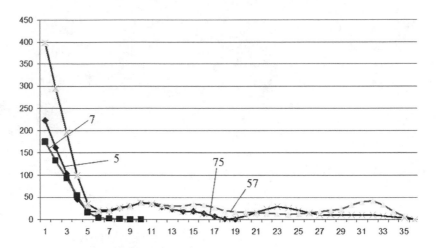

Fig. 8.42: Graphic representation of the FAI (45°) curves for images of symbols '5', '7' and images of numbers '57' and '75'

numbers '57' and '75' are equal to the sum of the FAI (135°) for the images of symbols '5', '7'. The duration of the FAI (135°) curve for the image of the symbol '5' is greater than that for the image of the symbol '7'.

The numerical values of the FAI (135°) for the images of symbols '5', '7' and images of the numbers '57' and '75' are shown in Table 8.12.

The shaded areas of Table 8.10 indicate in what time-steps the sum of the FAI (135°) for the images of symbols '5' and '7' is equal to the FAI (135°) for the images of numbers '57' and '75'. We can also determine which symbol is located first and which second. Table also allows us to determine the width of the space between the characters. The FAI (135°) curves for the images of the numbers '5', '7', '57' and '75' are shown in Fig. 8.43.

Another situation occurs when more than two characters are combined. However, they are also solved using all the directions of the shift. For example, the FAI (45°) for the images of the symbols 'a', 'n', 'd', 't', 'h', 'e' and images of the symbol groups 'and' and 'the' in Table 8.13 are shown. Also in Table 8.13, the values of the sums of the FAI (45°) of the individual symbols and the difference of these sums between the FAI (45°) symbol images are presented.

In Table 8.13, the symbols Σ_{and} and Σ_{the} are denoted by the sums of the FAI (45°) for the images of the symbols 'a', 'n', 'd' and 't', 'h', 'e' respectively. The shaded areas in Table 8.13 indicate the coincidence of the FAI (45°) sums during the shift in the 45° direction.

A similar Table 8.14 is presented for the FAI (135°) of the same images.

Table 8.12: Numerical values of the FAI (135°) for images of symbols '5', '7' and images of numbers '57' and '75'

Step number of the image shift	FAI (135°)				$\Sigma_{5,7}$	$\Sigma_{5,7}-$ FAI^{57} (135°)	$\Sigma_{5,7}-$ FAI^{75} (135°)
	Image of '5' character	Image of '7' character	Image of number '57'	Image of number '75'			
0	223	175	398	398	398	0	0
1	157	107	264	264	264	0	0
2	107	46	153	153	153	0	0
3	76	24	100	100	100	0	0
4	54	12	66	66	66	0	0
5	45	12	57	57	57	0	0
6	40	12	52	52	52	0	0
7	35	12	47	47	47	0	0
8	31	12	43	43	43	0	0
9	29	12	41	44	41	0	-3
10	27	12	39	46	39	0	-7
11	25	14	39	46	39	0	-7
12	24	17	41	46	41	0	-5
13	21	15	36	41	36	0	-5
14	16	7	23	32	23	0	-9
15	10	0	10	22	10	0	-12
16	2	0	4	13	2	-2	-11
17	0	0	3	10	0	-3	-10
18	0	0	3	9	0	-3	-9
19	0	0	4	9	0	-4	-9
20	0	0	8	8	0	-8	-8
21	0	0	13	8	0	-13	-8
22	0	0	15	8	0	-15	-8
23	0	0	13	15	0	-13	-15
24	0	0	12	19	0	-12	-19
25	0	0	11	16	0	-11	-16
26	0	0	10	15	0	-10	-15
27	0	0	11	21	0	-11	-21
28	0	0	12	23	0	-12	-23
29	0	0	13	29	0	-13	-29
30	0	0	6	29	0	-6	-29
31	0	0	0	23	0	0	-23
32	0	0	0	15	0	0	-15
33	0	0	0	5	0	0	-5
34	0	0	0	0	0	0	0

Table 8.13: Numerical values of the FAI (45°) for images of symbols 'a', 'n', 'd', 't', 'h', 'e' and images of symbol groups 'and' and 'the'

| Step number of the image shift | FAI (45°) | | | | | | | | Σ_{and} | $\Sigma_{and} - FAI_{and}(45°)$ | Σ_{the} | $\Sigma_{and} - FAI_{and}(45°)$ |
	Image of 'a' character	Image of 'n' character	Image of 'd' character	Image of 'and' symbol group	Image of 't' character	Image of 'h' character	Image of 'e' character	Image of 'the' symbol group				
0	96	117	135	348	70	140	75	285	348	0	285	0
1	65	80	93	238	49	96	45	190	238	0	190	0
2	40	51	60	151	31	61	25	117	151	0	117	0
3	29	30	34	93	16	35	15	66	93	0	66	0
4	22	21	17	66	4	20	12	42	60	-6	36	-6
5	22	23	20	78	2	23	14	54	65	-13	39	-15
6	23	23	30	96	1	23	16	63	76	-20	40	-23
7	17	21	34	100	0	21	12	60	72	-28	33	-27
8	10	14	35	90	0	14	3	52	59	-31	17	-35
9	4	8	32	72	0	8	0	47	44	-28	8	-39
10	0	4	20	48	0	4	0	46	24	-24	4	-42

(Contd.)

Table 8.13: *(Contd.)*

11	0	1	10	1	31	1	0	43	11	-20	1	-42
'11	0	1	10	1	31	1	0	43	11	-20	1	-42
12	0	0	3	0	14	0	0	24	3	-11	0	-24
13	0	0	0	0	9	0	0	9	0	-9	0	-9
14	0	0	0	0	13	0	0	3	0	-13	0	-3
15	0	0	0	0	14	0	0	0	0	-14	0	0
16	0	0	0	0	12	0	0	0	0	-12	0	0
17	0	0	0	0	7	0	0	0	0	-7	0	0
18	0	0	0	0	6	0	0	0	0	-6	0	0
19	0	0	0	0	0	0	0	0	0	0	0	0

Table 8.14: Numerical values of the FAI (135°) for images of symbols 'a', 'n', 'd', 't', 'h', 'e' and images of symbol groups 'and' and 'the'

| Step number of the image shift | FAI (135°) | | | | | | | | Σ_{and} | $\Sigma_{and} - FAI_{and}(13°)$ | Σ_{the} | $\Sigma_{and} - FAI_{and}(135°)$ |
	Image of 'a' character	Image of 'n' character	Image of 'd' character	Image of 'and' symbol group	Image of 't' character	Image of 'h' character	Image of 'e' character	Image of 'the' symbol group				
0	96	117	135	348	70	140	75	285	348	0	285	0
1	61	82	94	237	48	99	46	193	237	0	193	0
2	42	55	64	161	33	69	32	134	161	0	134	0
3	30	33	36	99	21	43	22	86	99	0	86	0
4	18	22	18	61	8	30	18	59	58	-3	56	-3
5	17	23	14	62	4	36	11	59	54	-8	51	-8
6	16	24	16	70	2	44	7	67	56	-14	53	-14
7	11	24	17	74	0	49	6	75	52	-22	55	-20
8	9	18	15	68	0	36	3	69	42	-26	39	-30
9	4	12	9	51	0	24	0	58	25	-26	24	-34

(Contd.)

Table 8.14: *(Contd.)*

	1	7	4	34	0	13	0	46	12	-22	13	-33
10	1	7	4	34	0	13	0	46	12	-22	13	-33
11	0	4	1	21	0	4	0	35	5	-16	4	-31
12	0	2	0	11	0	1	0	27	2	-9	1	-26
13	0	0	0	0	0	0	0	16	0	0	0	-16
14	0	0	0	0	0	0	0	16	0	0	0	-16
15	0	0	0	0	0	0	0	16	0	0	0	-16
16	0	0	0	0	0	0	0	13	0	0	0	-13
17	0	0	0	0	0	0	0	9	0	0	0	-9
18	0	0	0	0	0	0	0	5	0	0	0	-5
19	0	0	0	0	0	0	0	0	0	0	0	0

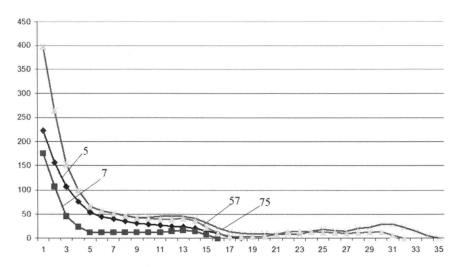

Fig. 8.43: Graphic representation of the FAI (135°) curves for the images of the symbols '5', '7' and images of numbers '57' and '75'

These examples of experimental studies are considered for small groups of symbols in order to better understand the issue. As can be seen the groups of symbols can be increased, which allows recognition and reading of large numbers.

Recognising large-sized text documents is one of the tasks where the use of parallel shift technology gives highest efficiency. There is a large number of recognition tasks where parallel-shift technology can be used.

Among other things, unlike other image processing and recognition methods, parallel shift technology claims to be versatile and can be used to process and understand a wide range of images.

References

Acharya, T., B.B. Bhattacharya, A. Bishnu, M.K. Kundu and Ch.A. Murthy. 2006. Computing the Euler Number of a Binary Image. United States Patent 7027649 B1. April 11.

Ahmed, N., T. Natarajan and K.R. Rao. 1974. Discrete Cisine Transform. *IEEE Trans on Computers*, 90-93.

Aho, A.V., J.E. Hopcroft and J.D. Ullman. 1983. Data Structures and Algorithms. Addison-Wesley, Reading, Massachusetts.

Aizenberg, I.N., N.N. Aizenberg and G.A. Krivosheev. 1999. Multi-valued and Universal Binary Neurons: Learning Algorithms, Applications to Image Processing and Recognition. *Lecture Notes in Artificial Intelligence – Machine Learning and Data Mining in Pattern Recognition*, 21-35.

Aizerman, M.A., E.M. Braverman and L.I. Rozonoer. 1970. The Method of Potential Functions in Machine Learning Theory. 'Nauka'.

Anil, K. and K. Jain. 2010. DataClustering: 50 Years Beyondk-Means. *Pattern Recognition Lett.*, Vol. 31: 651-666.

Ashenden, P.J. 1990. The VHDL Cookbook. Dept. Computer Science University of Adelaide South Australia.

Ashenden, P.J. 2001. The Designer's Guide to VHDL, 2nd ed., Morgan Kaufmann.

Ballard, D.H. 1981. Generalizing the Hough Transform to Detect Arbitrary Shapes. *Pattern Recognition*, Vol. 13. No. 2.

Barnsley, M.F. 1988. Fractals Everywhere. London: Academic Press Inc.

Belan, S. 2011. Specialized Cellular Structures for Image Contour Analysis. *Cybernetics and Systems Analysis*, 47(5): 695-704.

Belan, S. and N. Belan. 2013. Temporal-Impulse Description of Complex Image Based on Cellular Automata. *LNCS*, Vol. 7979, 291-295. Springer-Verlag Berlin, Heidelberg.

Belan, S. and S. Yuzhakov. 2013. A Homogenous Parameter Set for Image Recognition Based on Area. *Computer and Information Science*, Vol. 6, No. 2: 93-102, Published by Canadian Centre of Science and Education.

Belan, S. and N. Belan. 2012. Use of Cellular Automata to Create an Artificial System of Image Classification and Recognition. ACRI2012, *LNCS*, 7495: 483-493. Springer-Verlag Berlin Heidelberg.

Belan, S. and S. Yuzhakov. 2013. Machine Vision System Based on the Parallel Shift Technology and Multiple Image Analysis. *Computer and Information Science,* Vol. 6, No 4: 115-124, Published by Canadian Centre of Science and Education.

Belan, S.N. and R.L. Motornyuk. 2013. Extraction of Characteristic Features of Images with the Help of the Radon Transform and Its Hardware Implementation in Terms of Cellular Automata. *Cybernetics and Systems Analysis,* Vol. 49, Issue 1: 7-14.

Belhumeur, P.N., J.P. Hespanha and D.J. Kriegman. 1997. Eigenfaces vs Fisherfaces: Recognition Using Class Specific Linear Projection. *IEEE Transactions on Pattern Analysis and Machine Intelligence,* Vol. 19: 711-720.

Berklim, P. 2002. Survey of Clastering Data Mining Techniques. Accure Software.

Bilan, S. 2014. Models and Hardware Implementation of Methods of Pre-processing Images based on the Cellular Automata. *Advances in Image and Video Processing,* Vol. 2, No. 5: 76-90.

Bilan, S., R. Motornyuk and S. Bilan. 2014. Method of Hardware Selection of Characteristic Features Based on Radon Transformation and not Sensitive to Rotation, Shifting and Scale of the Input Images. *Advances in Image and Video Processing,* Vol. 2, No. 4: 12-23, UK.

Bilan, S., S. Yuzhakov and S. Bilan. 2014. Saving of Etalons in Image Processing Systems Based on the Parallel Shift Technology. Advances in Image and Video Processing – Vol 2, No 6: 36-41

Bishnu, B., B. Bhattacharya, M.K. Kundu, C.A. Murthy and T. Acharya. 2005. A Pipeline Architecture for Computing the Euler Number of a Binary Image. *Journal of Systems Architecture,* 51(8): 47-487.

Bracewell, R.N. 1984a. The Fast Hartley Transform. *Proc. IEEE,* Vol. 72, No. 8: 1010-1018.

Bracewell, R.N. 1984b. The Discrete Hartley Transform. *J. Opt. Am.,* Vol. 73, No. 12: 1832-1835.

Bracewell, R.N. 1986. The Fourier Transform and its Applications. Revised 2nd Edition. McGraw Hill Book Co, Singapore.

Chang, Ray-I, Shu-Yu Lin, Jan-Ming Ho, Chi-Wen Fann and Yu-Chun Wang. 2012. A Novel Content Based Image Retrieval System using K-means/KNN with Feature Extraction. *ComSIS,* Vol. 9, No. 4: 1645-1661. Special Issue, December.

Chen, C.H., L.F. Rau and P.S.P. Wang. 1995. Handbook of Pattern Recognition and Computer Vision. Singapore-New Jersey-London-Hong Kong: World Scientific Publishing Co. Pte. Ltd.

Chen, X. and A. Yuille. 2004. Detecting and Reading Text in Natural Scenes. *Computer Vision and Pattern Recognition* (CVPR): 366-373.

Dai, Y. and Y. Nakano. 1998. Recognition of Facial Images with Low Resolution Using a Hopfield Memory Model. *Pattern Recognition,* Vol. 31: 159-167.

Daubechies, I. 1990. The Wavelet Transform, Time Frequency Localisation and Signal Analysis. *IEEE Trans. on Information Theory,* Vol. 36, No. 5: 961-1004.

Duda, R.O. and P.E. Hart. 1973. Pattern Classification and Scene Analysis. NewYork: J. Wiley & Sons.

Duran, B.S. and P.L. Odell. 1974. Cluster Analysis: A Survey. Springer-Verlag, Berlin.

Epshtein, B., E. Ofek and Y. Wexler. 2012. Detecting Text in Natural Scenes with Stroke Width Transform – Microsoft Corporation, 2009. http://research.microsoft.com/pubs/149305/1509.pdf. — 2010

Foltyniewicz, R. 1995. Efficient High Order Neural Network for Rotation, Translation and Distance Invariant Recognition of Gray Scale Images. *Lecture Notes in Computer Science – Computer Analysis of Images and Patterns*, 424-431.

Fukushima, K. 1983. Neural Network for Visual Pattern Recognition. *Comput*, Vol. 21, No. 3: 65-115.

Gimel'farb, G.L., E.R. Hancock, A. Imiya, A. Kuijper, M. Kudo, S. Omachi, T. Windeatt and K. Yamada. 2012. Proceedings Joint IAPR International Workshop "Structural, Syntactic, and Statistical Pattern Recognition", SSPR&SPR 2012, Hiroshima, Japan, November 7-9.

González, Á. and L.M. Bergasa. 2013. A Text Reading Algorithm for Natural Images. *Image and Vision Computing*, Vol. 31, No. 3: 255-274.

Gonzalez, R.C. and R.E. Woods. 2008. Digital Image Processing. 3rd ed., Prentice Hall, New Jersey.

Gonzalez, R.C., R.E. Woods and S.L. Eddins. 2004. Digital Image Processing using MATLAB.

Hancock, E.R., R.C. Wilson, T. Windeatt, I. Ulusoy and F. Escolano. 2010. Proceedings Joint IAPR International Workshop "Structural, Syntactic, and Statistical Pattern Recognition", SSPR&SPR 2010, Cesme, Izmir, Turkey, August 18-20.

Heinzelman, W.B., A.P. Chandrakasan and H. Balakrishnan. 2002. An application specific protocol architecture for wireless microsensor networks. *IEEE Transactions on Wireless Communications*, Vol. 1, No. 4: 660-670.

ICANN. 2014. The 24th International Conference on Artificial Neural Networks, 15-19 September 2014. Hamburg, Germany.

Islam, S. and Dr. M. Ahmed. 2013. Implementation of Image Segmentation for Natural Images using Clustering Methods. *International Journal of Emerging Technology and Advanced Engineering*. Vol. 3, No. 3: 175-180.

Jacobsen, X., U. Zscherpel and P. Perner. 1999. A Comparison between Neural Networks and Decision Trees. *Lecture Notes in Artificial Intelligence – Machine Learning and Data Mining in Pattern Recognition*, 144-158.

Koh, E., D. Metaxas and N. Baldev. 1994. Hierarchical Shape Representation Using Locally Adaptive Finite Elements. *Lecture Notes in Computer Science*, No. 300: 441-446.

Konstantinos, B.B. 2011. Hexagonal is Circular Cell Shape: A Comparative Analysis and Evaluation of the Two Popular Modeling Approximations. *Cellular Networks, Positioning, Performance Analysis, Reliability*, 103-122.

Kozhemyako, V., S. Bilan and I. Savaliuk. 2001. Optoelectronic self-regulation neural system for treatment of vision information. *SPIE Proceedings*, 120-126. 3055, Wasington, USA.

Krzyzak, A., W. Dai and C.Y. Suen. 1990. Unconstrained Handwritten Character Classification Using Modified Backpropagation Model. *Proc. 1st Int. Workshop on Frontiers in Handwriting Recognition*, 155-166, Montreal, Canada.

Lawrence, S., C.L. Giles, A.C. Tsoi and A.D. Back. 1997. Face Recognition: A Convolutional Neural Network Approach. *IEEE Transactions on Neural Networks*, 1-24. Special Issue on Neural Networks and Pattern Recognition.

Lienhart, R. and A. Wernicke. 2002. Localizing and Segmenting Text in Images and Videos. *IEEE Transactions on Circuits and Systems for Video Technology*, Vol. 12, No. 4: 256-268.

Lu, Q., W. Luo, J. Wang and B. Chen. 2008. Low-complexity and energy efficient image compression scheme for wireless sensor networks. *Computer Networks*, Vol. 52, No. 13: 2594-2603.

Maier, A. 2012. Accelerated Centre-of-Gravity Calculation for Massive Numbers of Image Patches. *Advances in Visual Computing*, Vol. 7431: 566-574. Series Lecture Notes in Computer Science.

Mealy, Bryan and Tappero Fabrizio (2012). Free Range VHDL. Publisher: freerangefactory.org.

Milanova, M., P.E.M. Almeida, J. Okamoto and M.G. Simoes. 1999. Applications of Cellular Neural Networks for Shape from Shading Problem. *Lecture Notes in Artificial Intelligence – Machine Learning and Data Mining in Pattern Recognition*, 51-63.

Mills, R.L. 2006. Novel Method and System for Pattern Recognition and Processing Using Data Encoded as Fourier Series in Fourier Space. *Engineering Applications of Artificial Intelligence*, Vol. 19, Issue 2: 219-234.

Minsky, M. and S. Papert. 1969. Perceptrons: An Introduction to Computational Geometry. Cambridge, MA. MIT Press.

Narasimhan, R.N. 1966. Syntax-directed Interpretation of Classes of Pictures. *Comm. ACM*, No. 9: 166-173.

Nicoladie, D.T. 2014. Hexagonal Pixel-array for Efficient Spatial Computation for Motion-detection Pre-processing of Visual Scenes. *Advances in Image and Video Processing*, No. 2 (2): 26-36.

Nixon, M.S. and Alberto S. Aguardo. 2002. Feature Extraction and Image Processing. Newnes.

Parker, J.R. 2010. Algorithms for Image Processing and Computer Vision. Second Edition. Wiley Publishing, Inc.

Petrou, M. and J. Kittler. 1991. Optimal Edge Detectors for Ramp Edges. *IEEE Transactions on Pattern Analysis and Machine Intelligence*, Vol. 13(5): 483-491.

Pong, P. Chu. 2006. RTL Hardware Design Using VHDL: Coding for Efficiency, Portability, and Scalability. John Wiley & Sons, Inc.

Pratt, W.K. 2016. Digital Images Processing. Third edition. Wiley.

Ranganath, S. and K. Arun. 1997. Face Recognition Using Transform Features and Neural Networks. *Pattern Recognition*, Vol. 30: 1615-1622.

Rushton, A. 1998. VHDL for Logic Synthesis, 2nd ed., John Wiley & Sons.

Saupe, D., R. Hamzaoui and H. Hartenstein. 1996. Fractal Image Compression – An Introductory Overview. In: Fractal Models for Image Synthesis, Compression and Analysis, SIGGRAPH'96 Coure Notes, ACM, New Orleans, Louisiana. Aug.

Shalkoff, R.J. 1989. Digital Image Processing and Computer Vision. New York-Chichester-Brisbane-Toronto-Singapore: John Wiley & Sons, Inc.

Shaw, A.C. 1969. A Formal Picture Description Scheme as a Basis for Picture Processing System. *Information and Control*, No. 14: 9-52.

Simek, M., P. Moravek and J. Silva. 2011. Modeling of Energy Consumption of Zigbee Devices in Matlab Tool. *Elektrorevue Article*, Vol. 2, No. 3: 41-46.

Solomon, C. and T. Breckon. 2011. Fundamental of Digital Image Processing: A Practical Approach with Examples in Matlab. Wiley, Blackwell.

Sossa-Azuela, J.H., R. Santiago-Montero, M. Pérez-Cisneros and E. Rubio-Espino. 2013. Computing the Euler Number of a Binary Image Based on a Vertex Codification. *Journal of Applied Research and Technology*, Vol. 11: 360-370.

Spacek, L.A. 1986. Edge Detection of Contours and Motion Detection: Image Vision Compute, Vol. 4: 43-56.

Stallings, W.W. 1972. Recognition of Printed Chinese Characters by Automatic Pattern Analysis. *Comput. Graphics and Image Process.*, No. 1: 47-65.

Tahmasebi, P. and A. Harakhani. 2012. A Hybrid Neural Networks, Fuzzy Logic, Genetic Algorithm for Grade Estimation. *Computers & Geosciences*, Vol. 42: 18-27.

Tou, J.T. and R.C. Gonzalez. 1977. Pattern Recognition Principles. Addison-Wesley, New York, USA.

Tremeau, A. and N. Borel. 1997. Aregion Growing and Merging Algorithm to Color Segmentation: Pattern Recognition, 1191-1203.

Ulam, S. 1952. Random Processes and Transformations. *Procedings Int. Congr. Mathem.*, No. 2: 264-275.

Van Assen, H., M. Egmont-Petersen and J. Reiber. 2002. Accurate Object Localization in Gray Level Images Using the Centre of Gravity Measure: Accuracy Versus Precision. *IEEE Transactions on Image Processing*, 11: 1379-1384.

Vetter, T. and T. Poggio. 1997. Linear Object Classes and Image Synthesis from a Single Example Image. *IEEE Transactions on Pattern Analysis and Machine Intelligence*, Vol. 19: 733-742.

Walsh, D. and A.E. Raftery. 2002. Accurate and Efficient Curve Detection in Images: The Importance Sampling Hough Transform. *IEEE Transactions on Pattern Analysis and Machine Intelligence*, Vol. 10: 121-125.

Winston, P.H. 1992. Artificial Intelligence (3rd ed.). Addison-Wesley Longman Publishing Co., Inc. Boston, MA, USA.

Xingui, H. and X. Shaohua. 2010. Process Neural Networks: Theory and Application. Advanced Topics in Science and Technology in China.

Yan, J. and X. Gao. 2012. Detection and Recognition of Text Superimposed in Images base on Layered Method. *Neurocomputing*, Vol. 134: 3-14.

Yoon, K.S., Y.K. Ham and R.H. Park. 1998. Hybrid Approaches to Frontal View Face Recognition Using the Hidden Markov Model and Neural Network. *Pattern Recognition*, Vol. 31: 283-293.

Yuen, H., J. Illingworth and J. Kittler. 1989. Detecting Partially Occluded Ellipses Using the Hough Transform. *Image and Vision Computing*, Vol. 8, No. 1: 71-77.

Zimmermann, R. 1998. VHDL Library of Arithmetic Units. First International Forum on Design Languages (FDL'98).

Index

A

additional coefficients, 29, 74
additional parameters, 25, 28
area of intersection detection block, 101

B

basic parameters, 18, 19, 21, 22, 23, 25, 26, 29, 30, 33, 34, 44, 51, 53, 59, 74, 79, 91
binarisation, 36, 36, 37, 49, 66, 67, 128
bitmap images, 2, 3

C

calculated analytically, 27
cell, 2, 18, 40, 45, 48, 87, 102, 104, 105, 106, 107, 108, 109, 110, 112, 127
characteristic features, 5, 6, 7, 8, 9, 11, 18, 25, 30, 126, 127, 128, 166, 167, 168
cluster, 6, 8, 11, 12
combinational circuit, 106, 108, 111, 112
comparing unit, 102
comparison, 11, 25, 28, 29, 30, 31, 32, 36, 46, 47, 53, 54, 63, 72, 73, 82, 84, 85, 91, 93, 94, 97, 98, 101, 102, 113, 117, 118, 119, 128, 164
complex images, 2, 3, 4, 127, 166
contour, 3, 16, 28, 37, 39, 41, 42, 46, 51, 85, 90, 163, 164, 165
controlled power element, 109

counting the number of ones circuit, 111
current-code converter, 110
curvature, 5, 118, 129, 130, 149, 150, 151, 152, 153, 154, 155, 156, 159, 160, 164, 165
cyclic shifts, 57

D

density coefficient, 43
detailed stage, 47, 53, 85, 93
D-flip-flop, 106, 108, 115,
distance, 5, 8, 10, 11, 14, 15, 18, 20, 33, 38, 51, 57, 58, 60, 61, 62, 63, 70, 121, 126, 130
dynamic objects, 7

E

edge detection, 4, 35, 36, 37, 38, 39, 40, 79, 166

F

features extracted, 6
function of the area of intersection, 13, 14, 15, 16, 18, 19, 20, 21, 22, 23, 24, 25, 26, 29, 30, 32, 42, 43, 44, 46, 47, 48, 49, 53, 55, 56, 57, 61, 62, 64, 66, 67, 70, 71, 72, 73, 74, 77, 82, 83, 84, 85, 91, 93, 97, 101, 103, 105, 113, 116, 117, 118, 122, 123, 124, 128, 130, 131, 132, 133, 141, 142, 146, 148, 149, 150, 151, 152, 155, 157,

158, 159, 161, 162, 163, 164, 165, 169, 170, 173, 176
function of the volume of intersection, 33, 34

H

handwritten text, 166

I

image area, 7, 18, 42, 44, 45, 47, 73, 74, 80, 96, 98
image copy, 18, 38, 39, 47, 56, 57, 59, 63, 70, 80
image noise, 37, 99
image processing, 3, 7, 12, 14, 35, 36, 37, 38, 40, 46, 49, 51, 56, 57, 64, 65, 66, 68, 69, 70, 71, 72, 73, 74, 77, 78, 80, 84, 85, 90, 120, 126, 131, 166, 186
image recognition, 2, 3, 4, 5, 6, 26, 28, 29, 30, 31, 32, 47, 52, 82, 83, 84, 58, 87, 90, 91, 93, 94, 96, 98, 99, 100, 101, 104, 113, 120, 127, 130, 131, 163, 165, 166, 167, 168
image transformation, 6, 7, 122, 127
images of the symbols, 168, 170, 171, 177, 178
integral coefficient, 26, 27, 28, 29, 30, 32, 33, 34, 43, 53, 74, 82, 84, 91, 92
intelligent systems, 2, 50, 51, 65, 66, 68

M

machine vision system, 79
machine vision systems, 38, 50, 56, 64, 71
mass centre, 60, 61, 63
matches, 31, 32, 167, 177
matrices, 48, 90, 105, 160, 109
maximum shift, 15, 18, 25, 26, 28, 33, 44, 47, 57, 62, 74, 79, 85, 90
mirror of initial image, 120
mirrors of copy image, 120
motion parameters, 51, 55, 57, 58, 59, 63

N

neighbourhood, 3
non-bitmap images, 2, 3

non-cyclic Shifts, 55, 59
number of distortions, 134, 136, 138, 144

O

optical system, 3, 7, 62, 101, 120, 122
optical transparencies, 120, 123
orthogonal directions, 16, 20, 33, 38, 39, 57, 70, 84, 93, 97

P

parallel shift technology, 7, 8, 9, 11, 12, 14, 16, 29, 30, 33, 34, 36, 37, 40, 46, 47, 49, 50, 51, 52, 55, 57, 60, 64, 65, 66, 71, 72, 74, 77, 78, 79, 80, 82, 85, 90, 94, 96, 98, 99, 101, 113, 120, 126, 128, 131, 157, 160, 163, 165, 166, 168, 186
parallel shift, 7, 8, 9, 11, 12, 13, 14, 16, 18, 22, 29, 30, 32, 33, 34, 36, 37, 39, 40, 45, 46, 47, 49, 50, 51, 52, 55, 57, 60, 63, 64, 65, 66, 71, 72, 74, 77, 78, 79, 80, 82, 85, 90, 94, 96, 97, 98, 99, 101, 104, 113, 117, 120, 126, 128, 131, 157, 160, 163, 165, 166, 168, 169, 186
pattern surface, 72, 74, 79, 80, 82
pattern, 2, 6, 7, 29, 42, 52, 53, 54, 55, 70, 72, 73, 74, 76, 77, 79, 80, 82, 84, 91, 92, 93, 96, 97, 100, 117, 146, 157, 158, 160, 161, 162, 164, 165, 166, 167, 168
permissible error, 46, 47, 84, 87, 90, 91, 93, 94, 96
photodetector, 120, 121, 123

R

rectangle, 2, 18, 21, 22, 24, 25, 27, 28, 29, 94, 95, 96, 152
reverse shift registers, 107
rotation, 23, 24, 32, 48, 84, 97, 100, 122

S

scaling coefficient, 42, 43
shift directions, 11, 12, 39, 97, 102, 104, 132, 133, 159

spatial orientation, 28, 29, 52, 55, 70, 72, 77
square, 8, 18, 22, 27, 28, 44, 45, 55, 84, 85, 86, 94, 95, 98, 132, 134, 149, 151, 152, 155

T

template functions, 72, 102, 128
template images, 46, 72, 96, 99, 119
the memory block of the function of the area of intersecting etalons, 102

trajectory, 36, 64, 166
triangle, 2, 18, 22, 23, 24, 27, 28, 55, 94, 95, 104, 132, 143, 148, 149, 150, 151, 165

V

vector of the characteristic features, 6, 7
VHDL codes, 114